T0338267

# Small-Angle Scattering

## Professor Ian Hamley's Book Publications

Authored

*The Physics of Block Copolymers*
*Introduction to Soft Matter* (first and revised edition)
*Block Copolymers in Solution*
*Introduction to Peptide Science*

Edited

*Developments in Block Copolymer Science and Technology*
*Nanoscale Science and Technology* (with R.W. Kelsall and M. Geoghegan)
*Hydrogels in Cell-Based Therapies* (with C. Connon)

# Small-Angle Scattering

*Theory, Instrumentation, Data, and Applications*

IAN W. HAMLEY
*School of Chemistry, University of Reading, UK*

WILEY

This edition first published 2021
© 2021 John Wiley & Sons Ltd

The right of Ian W. Hamley to be identified as the author of this work has been asserted in accordance with law.

*Registered Offices*
John Wiley & Sons, Inc., 111 River Street, Hoboken, NJ 07030, USA
John Wiley & Sons Ltd, The Atrium, Southern Gate, Chichester, West Sussex, PO19 8SQ, UK

*Editorial Office*
The Atrium, Southern Gate, Chichester, West Sussex, PO19 8SQ, UK

For details of our global editorial offices, customer services, and more information about Wiley products visit us at www.wiley.com.

Wiley also publishes its books in a variety of electronic formats and by print-on-demand. Some content that appears in standard print versions of this book may not be available in other formats.

*Library of Congress Cataloging-in-Publication Data*

Name: Hamley, Ian W., author.
Title: Small-angle scattering : theory, instrumentation, data and
    applications / Ian W. Hamley.
Description: First edition. | Hoboken, NJ : Wiley, 2021. | Includes
    bibliographical references and index.
Identifiers: LCCN 2020043351 (print) | LCCN 2020043352 (ebook) | ISBN
    9781119768302 (hardback) | ISBN 9781119768333 (adobe pdf) | ISBN
    9781119768340 (epub)
Subjects: LCSH: Small-angle scattering.
Classification: LCC QC482 .H365 2021 (print) | LCC QC482 (ebook) | DDC
    537.5/3–dc23
LC record available at https://lccn.loc.gov/2020043351
LC ebook record available at https://lccn.loc.gov/2020043352

Cover Design: Wiley
Cover Image: GiroScience/Alamy Stock Photo

Set in 10/13pt SabonLTStd by SPi Global, Chennai, India

SKY9B34F1EF-23D9-44CC-B832-40A40D5499AB_031621

# Contents

# Preface

This book aims to provide an up-to-date and comprehensive account of small-angle scattering, both small-angle x-ray and small-angle neutron scattering. It discusses both the underlying theory as well as giving practical information and useful examples. The book aims to complement the handful of existing texts in the field but has a broader coverage, not being restricted solely to biological macromolecules or polymers or soft matter. The text is intended to serve two uses. First, it is a 'go-to' reference text as a source of detailed information and essential references for those already working in the field. Second, it should serve as a useful general introduction to the field for the non-expert. The writing of this text relies on more than 30 years of experience working across the field on many systems and in numerous types of small-angle scattering experiment, leading teams using many instruments across most major European facilities, as well as lab instruments.

I thank the Synchrotron Radiation Source, Daresbury, for hosting me as a visiting fellow in 2004 and Diamond Light Source for the joint appointment (with the University of Reading) 2005–2010. I would like to thank my PhD supervisors (back in the mists of time) at the University of Southampton, Prof. Geoffrey Luckhurst and Prof. John Seddon, for introducing me to the world of small-angle scattering. I would like to thank many people with whom I have worked at synchrotron and neutron facilities over the years. The following is an incomplete list (sorry for omissions):

At Risø: Jan Skov Pedersen, Kell Mortensen, Martin Vigild, Wim de Jeu (AMOLF, The Netherlands) and Frank Bates (University of Minnesota, USA), and Jan at Aarhus as well and Frank at NIST also. At SRS Daresbury: Anthony Gleeson, Günter Grossmann, Ernie (Bernd) Komanschek, Liz Towns-Andrews, Chris Martin, Wim Bras, Tony Ryan, Greg Diakun and Nick Terrill, and many members of my group when at the University of Leeds, especially John Pople and Patrick Fairclough and Tim Lodge (University of Minnesota) for SAS-related collaboration. At ISIS: Steve King,

Richard Heenan, Sarah Rogers, Ann Terry, and James Doutch. At LURE: Claudie Bourgaux. At LPS Université Paris-sud: Marianne Imperor-Clerc and Patrick Davidson. At ELETTRA: Heinz Amenitsch. At LLB: Laurence Noirez. At MLZ, Munich: Henrich Frielinghaus and at Oak Ridge National Lab: Bill Hamilton and George Wignall. At the ESRF: Olivier Diat, Pierre Panine, Narayan (Theyencheri Narayanan), Kristina Kvashnina, Daniel Hermida-Merino, Giuseppe Portale, Petra Pernot, Martha Brennich, Mark Tully, Adam Round, Gemma Newby, and Tom Arnold. At DESY: Sergio Funari and Dmitri Svergun. At the ILL: Peter Lindner and Lionel Porcar. At MaxLab: Tomás Plivelic and at PSI-Swiss Neutron Source: Joachim Kohlbrecher. At ALBA: Marc Malfois and at SOLEIL: Javier Perez. At Diamond: Nick Terrill, Katsuaki Inoue, Nathan Cowieson, Nikul Khunti, Charlotte Edwards-Gayle, and Rob Rambo. I would also like to especially thank Narayanan Theyencheri (Narayan) from the ESRF for a critical reading and valuable comments on the text. As usual, I take responsibility for any remaining errors and omissions. Biggest thanks go to Valeria Castelletto, who has been a team member/leader at many beamtime sessions – and (perhaps more importantly!) we have also shared our lives for the last 20 years, 'and it doesn't seem a day too long'.

Ian W. Hamley
*University of Reading, UK, 2020*

# 1

# Basic Theory

## 1.1  INTRODUCTION

Small-angle scattering (SAS) is an important technique in the characterization of the structure and order in nanostructured materials as well as biomolecules and other solutions and suspensions. This book covers both small-angle x-ray scattering (SAXS, the subject of Chapter 4) and small-angle neutron scattering (SANS, discussed in Chapter 5) as well as grazing incidence small-angle scattering (GISAS, Chapter 6). This book does not discuss small-angle light scattering (also known as static light scattering, SLS), which is a separate topic. Although there are many similarities in the theory, light scattering is the subject of many specialist texts [1, 2], as well as chapters in texts about general SAS [3, 4]. This book also includes in Chapters 3 and 4 discussion of wide-angle scattering, especially wide-angle x-ray scattering (WAXS), which can be performed along with SAXS in the characterization of certain nanomaterials including polymers and nanoparticle systems with crystal or partially crystalline ordering. Instrumentation for the different types of measurement is discussed in Chapter 3 and data analysis processes are discussed in Chapter 2.

SAS, by the nature of reciprocal space, is suited to probe structures with sizes in the approximate range 1–100 nm, which is the structural size scale corresponding to many types of soft and hard nanomaterial as well as biomolecules such as proteins in solution. Considering Bragg's law, the scattering from such large structures will be observed at small angles (less than a few degrees of scattering angle $2\theta$). Wide-angle scattering covers the 0.1–1 nm range. Ultra-small-angle scattering (USAXS and USANS), also

*Small-Angle Scattering: Theory, Instrumentation, Data and Applications,*
First Edition. Ian W. Hamley.
© 2021 John Wiley & Sons Ltd. Published 2021 by John Wiley & Sons Ltd.

discussed in this book (see e.g. Chapters 3 and 5), can extend up to 1000 nm or more, which overlaps with the size scale probed by light scattering.

SANS and SAXS have complementary characteristics (Section 5.14), which are discussed in the respective chapters (Chapters 3 and 4) dedicated to these methods. These arise from the distinct natures of neutrons and x-rays. Neutrons are nuclear particles, with a mass $1.675 \times 10^{-27}$ kg. They have spin half (i.e. they are fermions) and a finite magnetic moment $\mu_n = -9.662 \times 10^{-27}$ J T$^{-1}$, and zero charge. In contrast, x-rays are photons with spin 1 (they are bosons), no mass, and no magnetic moment. X-rays are a type of electromagnetic radiation with wavelengths in the approximate range 0.01–1 nm with overlap with gamma rays at short wavelengths and the extreme ultraviolet at long wavelengths. Despite the different nature of neutrons and x-rays, both exhibit wave-like diffraction by matter. Using de Broglie's relationship, the associated wavelength of neutrons (this is discussed quantitatively in Section 3.6, in terms of the velocity distribution of neutrons produced by reactor and spallation sources) can be calculated.

This chapter provides a summary of the theory that underpins SAS, starting from the basic equations for the wavenumber and scattering amplitude (Section 1.2). Section 1.3 introduces the essential theory concerning the scattered intensity and its relationship to real space correlation functions, for both isotropic and anisotropic systems. Section 1.4 discusses the Guinier approximation, often used as a first analytical technique to obtain the radius of gyration from SAS data. The separation of a SAS intensity profile into intra-molecular and inter-molecular scattering components, respectively termed form and structure factor is discussed in Section 1.5. These terms are discussed in more detail, Section 1.6 first considering different commonly used structure factors, then Section 1.7 focusses on examples of form factors and the effects of polydispersity on form factors. Form and structure factors for polymers are the subject of Section 1.8.


## 1.2   WAVENUMBER AND SCATTERING AMPLITUDE

In a SAS experiment, the intensity of scattered radiation (x-rays or neutrons) is measured as a function of angle and is presented in terms of wavenumber $q$. This removes the dependence on wavelength $\lambda$ which would change the scale in a plot against angle, i.e. SAS data taken at different wavelengths will superpose when plotted against $q$, this is useful for example on beamlines where data is measured at different wavelengths (this is more common with neutron beamlines). The wavenumber quantity is sometimes denoted $Q$ although in this book $q$ is used consistently. The difference between incident and diffracted wavevectors $\mathbf{q} = \mathbf{k}_s - \mathbf{k}_i$ and since $|\mathbf{k}| = \frac{2\pi}{\lambda}$, and the scattering

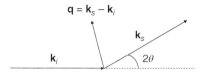

**Figure 1.1** Definition of wavevector **q** and scattering angle $2\theta$, related to the wavevectors of incident and scattered waves, $\mathbf{k_i}$ and $\mathbf{k_f}$.

angle is defined as $2\theta$ (Figure 1.1), the magnitude of the wavevector is given by

$$q = \frac{4\pi \sin\theta}{\lambda} \tag{1.1}$$

In some older texts, related quantities denoted $s$ or $S$ are used (these can correspond to $q/2$ or $q/2\pi$; the definition should be checked). The wavenumber $q$ has SI units of $nm^{-1}$, although $\text{Å}^{-1}$ is commonly employed.

The amplitude of a plane wave scattered by an ensemble of $N$ particles is given by

$$A(\mathbf{q}) = \sum_{j=1}^{N} a_j \exp[-i\mathbf{q}.\mathbf{r}_j] \tag{1.2}$$

Here, the scattering factors $a_j$ are either the ($q$-dependent) atomic scattering factors $f_j(q)$ (Section 4.4) for SAXS or the $q$-independent neutron scattering lengths $b_j$ for SANS (Section 5.4).

For a continuous distribution of scattering density, Eq. (1.2) becomes

$$A(\mathbf{q}) = \int \Delta\rho(\mathbf{r}) \exp[-i\mathbf{q}.\mathbf{r}]d\mathbf{r} \tag{1.3}$$

Here $\Delta\rho(\mathbf{r})$ is the excess scattering density above that of the background (usually solvent) scattering, which is a relative electron density in the case of SAXS or a neutron scattering length density (Eq. (5.11)) in the case of SANS.

## 1.3 INTENSITY FOR ANISOTROPIC AND ISOTROPIC SYSTEMS AND RELATIONSHIPS TO PAIR DISTANCE DISTRIBUTION AND AUTOCORRELATION FUNCTIONS

### 1.3.1 General (Anisotropic) Scattering

In the following, notation to indicate that the intensity is ensemble or time-averaged is not included for convenience (if the system is ergodic, which is often the case apart from certain gels and glasses etc., these two averages are equivalent).

The intensity is defined as

$$I(\mathbf{q}) = A(\mathbf{q})A^*(\mathbf{q}) \tag{1.4}$$

Thus, using Eq. (1.2), for an ensemble of discrete scattering centres

$$I(\mathbf{q}) = \sum_{j=1}^{N} \sum_{k=1}^{N} a_j a_k \exp[-i\mathbf{q}.(\mathbf{r}_j - \mathbf{r}_k)] \tag{1.5}$$

Whereas, for a continuous distribution of scattering density,

$$I(\mathbf{q}) = \int \int \Delta\rho(\mathbf{r}')\Delta\rho(\mathbf{r}'') \exp[-i\mathbf{q}.(\mathbf{r}' - \mathbf{r}'')]d\mathbf{r}' d\mathbf{r}'' \tag{1.6}$$

Equation (1.6) can also be rewritten in terms of an autocorrelation function (sometimes known as convolution square function) writing $\mathbf{r}' - \mathbf{r}'' = \mathbf{r}$

$$\gamma(\mathbf{r}) = \int \Delta\rho(\mathbf{r} + \mathbf{r}'')\Delta\rho(\mathbf{r}'')d\mathbf{r}'' \tag{1.7}$$

Then

$$I(\mathbf{q}) = \int \gamma(\mathbf{r}) \exp[-i\mathbf{q}.\mathbf{r}]d\mathbf{r} \tag{1.8}$$

The autocorrelation function has the physical meaning of the overlap between a particle and its 'ghost particle' displaced by r (Figure 1.2). This function is the continuous version of the Patterson function familiar from crystallography.

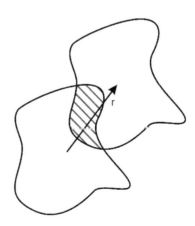

**Figure 1.2**   Ghost particle construction. The overlap volume (shaded) is the auto-correlation function.

The autocorrelation function for solid geometrical bodies can be calculated analytically. For a sphere of radius $R$ the expression is isotropic and is given by [5–7]

$$\gamma(r) = 1 - \frac{3}{2}\left(\frac{r}{2R}\right) + \frac{1}{2}\left(\frac{r}{2R}\right)^3 \tag{1.9}$$

This is a smoothly decaying function of $r$. The expression for a cylinder is provided in Ref. [8] and can be calculated for other structures, asymptotic expressions for cylinders and discs are given in Eqs. (1.83) and (1.84).

Equation (1.6) can alternatively be written for uncorrelated scatterers as

$$I(\mathbf{q}) = \left| \int \Delta\rho(\mathbf{r})\exp[-i\mathbf{q}.\mathbf{r}]d\mathbf{r} \right|^2 \tag{1.10}$$

## 1.3.2 Isotropic Scattering Systems

For isotropic scattering the scattered intensity will only be a function of the wavenumber $q$ and an orientational average (indicated by $<..>_\Omega$) is performed, i.e. Eq. (1.5) becomes

$$I(q) = \sum_{j=1}^{N}\sum_{k=1}^{N} a_j a_k < \exp[-i\mathbf{q}.\mathbf{r}_{jk}]>_\Omega \tag{1.11}$$

where $\mathbf{r}_{jk} = \mathbf{r}_j - \mathbf{r}_k$ .

The average over all orientations of $\mathbf{r}_{jk}$ can be evaluated as follows

$$< \exp[-i\mathbf{q}.\mathbf{r}_{jk}]>_\Omega = \frac{1}{4\pi}\int_0^{2\pi} d\phi \int_0^\pi \exp(-iqr_{jk}\cos\theta)\sin\theta d\theta = \frac{\sin(qr_{jk})}{qr_{jk}} \tag{1.12}$$

This leads to the Debye equation for scattering from an isotropically averaged ensemble:

$$I(q) = \sum_{j=1}^{N}\sum_{k=1}^{N} a_j a_k \frac{\sin(qr_{jk})}{qr_{jk}} \tag{1.13}$$

Considering a continuous distribution of scattering density, the orientational averaging of Eq. (1.12) has to be performed over $\Delta\rho(\mathbf{r})$ since it is a function of $\mathbf{r}$:

$$I(\mathbf{q}) = \left\langle \int\int \Delta\rho(\mathbf{r}')\Delta\rho(\mathbf{r}'')\exp[-i\mathbf{q}.(\mathbf{r}'-\mathbf{r}'')]d\mathbf{r}'d\mathbf{r}'' \right\rangle_\Omega \tag{1.14}$$

The isotropic average of Eq. (1.14) leads, via Eq. (1.12), to

$$I(q) = 4\pi \int_0^{D_{max}} \Delta\rho^2(r)\frac{\sin(qr)}{qr}r^2 dr \tag{1.15}$$

Here, $D_{max}$ is the maximum dimension of a particle (maximum distance from the geometric centre).

In terms of the isotropically averaged autocorrelation function $\gamma(r)$ this can be written as

$$I(q) = 4\pi \int_0^{D_{max}} \gamma^2(r) r^2 \frac{\sin(qr)}{qr} dr \tag{1.16}$$

Or in terms of the Debye correlation function $\Gamma(r)$:

$$I(q) = 4\pi \int_0^{D_{max}} \Gamma(r) r^2 \frac{\sin(qr)}{qr} dr \tag{1.17}$$

This then leads to the expression

$$I(q) = 4\pi \int_0^{D_{max}} p(r) \frac{\sin(qr)}{qr} dr \tag{1.18}$$

Here $p(r) = \Gamma(r) r^2$ is the pair distance distribution function (PDDF). This is an important quantity in SAS data analysis since as can be seen from Eq. (1.18), it is related to the intensity via an indirect Fourier transform:

$$p(r) = \frac{1}{2\pi^2} \int_0^{q_{max}} I(q) qr \sin(qr) dq \tag{1.19}$$

The PDDF provides information on the shape of particles, as well as their maximum dimension $D_{max}$. Figure 1.3 compares the PDDF of different shaped objects.

Many SAS data analysis software packages such as ATSAS and others (Table 2.2) and software on synchrotron beamlines is able to compute PDDFs from measured data. Methods to obtain PDDFs by indirect Fourier transform methods are discussed further in Section 4.6.1.

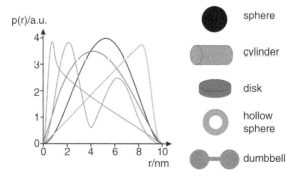

**Figure 1.3**   Sketches of pair distance distribution functions for the colour-coded particle shapes shown with a $D_{max} = 10$ nm. *Source*: Adapted from Ref. [9].

The radius of gyration can be obtained from $p(r)$ via the second moment [4, 7, 10, 11]:

$$R_g^2 = \frac{\int_0^{D_{max}} p(r)r^2\,dr}{2\int_0^{D_{max}} p(r)\,dr} \tag{1.20}$$

## 1.4   GUINIER APPROXIMATION

The Guinier equation is used to obtain the radius of gyration from a simple analysis of the scattering at very low $q$ (from the first part of the measured SAS intensity profile). The Guinier approximation can be obtained from Eq. (1.18), substituting the expansion [6, 7, 11]:

$$\frac{\sin(qr)}{qr} = 1 - \frac{(qr)^2}{3!} + \frac{(qr)^4}{5!} - \cdots \tag{1.21}$$

gives at sufficiently low $q$ (such that the expansion can be truncated at the second term)

$$I(q) = 4\pi \int_0^{D_{max}} p(r)\left(1 - \frac{(qr)^2}{3!}\right)dr \tag{1.22}$$

Considering the expression for the radius of gyration as the second moment of $p(r)$ (Eq. (1.20)) we obtain [6, 11, 12]

$$I(q) = I(0)\left(1 - \frac{q^2 R_g^2}{3} + \cdots\right) \tag{1.23}$$

Using the series expansion $e^x = 1 + x + \frac{x^2}{2!} + \cdots$ with $x = q^2 R_g^2$, and truncating at the second term (valid if $q$ is small), this can be rewritten as an exponential

$$I(q) = I(0)\exp\left(-\frac{q^2 R_g^2}{3}\right) \tag{1.24}$$

This is the Guinier equation. A Guinier plot of $\ln I(q)$ vs $q^2$ has slope $R_g^2/3$ at low $q$. Figure 2.8 shows representative Guinier plots. The Guinier equation (Eq. (1.23)) can also be obtained starting from Eq. (1.10), using the same series expansion for the exponential $\exp[-i\mathbf{q}.\mathbf{r}]$.

For a homogeneous sphere of radius $R$, the radius of gyration is given by $R_g = R\sqrt{3/5}$ [6, 11] whereas for a homogeneous infinite cylinder of radius $R$ it is given by $R_g = R/\sqrt{2}$ and for a thin disc of thickness $T$, $R_g = T/\sqrt{12}$ [6, 7, 10, 13]. For an ellipse with semiaxes $a$ and $b$, $R_g = \sqrt{a^2 + b^2}/2$. For a rod of length $L$ with finite cross-section the overall radius of gyration, $R_g$, is related to that of the cross-section $R_c$ via the expression [7, 11, 12, 14]:

$$R_g^2 = R_c^2 + \frac{L^2}{12} \tag{1.25}$$

The Guinier approximation is useful for systems containing non-interacting particles (i.e. the structure factor $S(q) = 1$, see Section 1.5) and is typically valid for $q < 1/R_g$.

## 1.5   FORM AND STRUCTURE FACTORS

The total intensity scattered by an ensemble of particles (self-assembled structures, surfactant or polymer assemblies, colloids, etc.) can be separated into terms depending on intra-particle and inter-particle terms (Figure 1.4), which are termed the *form factor* and *structure factor*, respectively.

For a monodisperse system of spherically symmetric particles (number density $n_p = N/V$), the scattering can be written as

$$I(q) = \sum_j \sum_k F_j(q) F_k(q) \frac{\sin(qr_{jk})}{qr_{jk}} \tag{1.26}$$

where $F(q)$ is the amplitude of scattering from within a particle:

$$F(q) = \left\langle \sum_i a_i \exp[-i\mathbf{q}.\mathbf{r}] \right\rangle_\Omega \tag{1.27}$$

which is analogous to an isotropic average over Eq. (1.2).

Equation (1.25) can be separated into intra- and inter- particle components:

$$I(q) = \sum_j F_j^2(q) + \sum_{j \neq k} \sum_k F_j(q) F_k(q) \frac{\sin(qr_{jk})}{qr_{jk}} \tag{1.28}$$

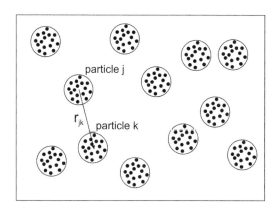

**Figure 1.4**   Scattering within particles (from atoms/components shown as black dots) corresponds to form factor while scattering between particles (separated by vector $\mathbf{r}_{jk}$) corresponds to structure factor.

This can be written for the monodisperse ensemble of spherically symmetric particles as

$$I(q) = P(q)S(q) \tag{1.29}$$

Here the form factor is

$$P(q) = \sum_j F_j^2(q) \tag{1.30}$$

and the structure factor is

$$S(q) = 1 + \sum_{j \neq k} \sum_k F_j(q)F_k(q)\frac{\sin(qr_{jk})}{qr_{jk}} \tag{1.31}$$

In a sufficiently dilute system only the form factor needs to be considered. Intermolecular interferences are manifested by the increasing contribution of the structure factor as concentration is increased. In many micellar, biomolecular, and surfactant systems in dilute aqueous solution, structure factor effects may not be observed (over the typical $q$ range accessed in most SAXS measurements). At high concentration, the structure factor is characterized by a series of peaks (due to successive nearest neighbour correlations, next nearest neighbour correlations etc.) the intensity decaying as $q$ increases, oscillating around the average value $S(q) = 1$. The structure factor is related by a Fourier transform to the radial distribution function $g(r)$:

$$S(q) = 1 + \frac{4\pi N}{V} \int_0^\infty g(r)\frac{\sin(qr)}{qr}r^2 dr \tag{1.32}$$

where $V$ is the volume of the particle.

Figure 1.5 shows the calculated total intensity within the monodisperse approximation (product of form and structure factors) for spheres of radius $R = 30$ Å (also hard sphere structure factor radius $R_H = 30$ Å) at two volume fractions showing the increase in the contribution from the structure factor at low $q$ including the development of a peak at $q \approx 0.1$ Å$^{-1}$.

The preceding derivation (Eqs. (1.26)–(1.32)) applies for the case of monodisperse particles with spherical symmetry. Other expressions have been introduced for other cases. For particles that are slightly anisotropic, the 'decoupling' approximation [16] is often used. The intensity for monodisperse particles is written as

$$I(q) = P(q)[1 + \beta(q)(S(q) - 1)] \tag{1.33}$$

where

$$\beta(q) = \langle F(q) \rangle^2 / \langle F^2(q) \rangle \tag{1.34}$$

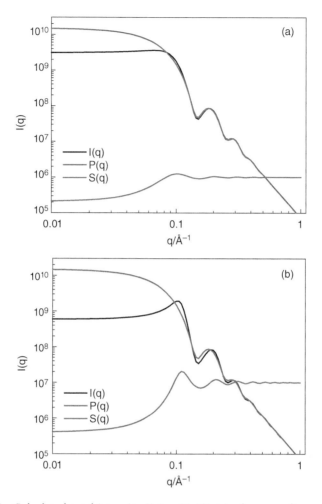

**Figure 1.5** Calculated total intensity $I(q) = P(q)S(q)$ in the monodisperse approximation for spheres with radius $R = 30\,\text{Å}$ (polydispersity $\sigma = 3\,\text{Å}$, see Figure 1.19) and hard sphere structure factor with $R_{HS} = 30\,\text{Å}$ and volume fraction (a) 0.2, (b) 0.4. The structure factor $S(q)$ has been scaled by a factor of $10^5$ in (a) and $10^6$ in (b) for ease of visualisation.

Here, $F(q)$ is the 'amplitude' form factor, given by Eq. (1.27). Note that there is confusion in the literature, and the term structure factor can be used to refer to intensity or amplitude according to the context. Here, $S(q)$ is used for intensity structure factor and $F(q)$ for form amplitude factor. In Eq. (1.34), the structure factor is calculated for the average particle size defined as $R_{av} = [3\,V/(4\pi)]^{1/3}$ [17].

For systems with small polydispersities, a decoupling approximation can also be used [17] according to the expression:

$$I(q) = \langle P(q) \rangle [1 + \beta(q)(S(q) - 1)] \tag{1.35}$$

where

$$\beta(q) = \left[ \int D(R)F(q, R)dR \right]^2 \Bigg/ \left[ \int D(R)F^2(q, R)dR \right] \tag{1.36}$$

Here, $D(R)$ is the dispersity distribution function (Section 1.7.4), which may have a Gaussian or log-normal function form, for example. The structure factor $S(q)$ is evaluated for the average particle size.

An alternative approximation is the local monodisperse approximation, which treats the system as an ensemble of locally monodisperse components (i.e. a particle of a given size is surrounded by particles of the same size). The intensity is then [17]

$$I(q) = \int D(R)F^2(q, R)S(q, R)dR \tag{1.37}$$

Other approximations for the factoring of form and structure factors have been proposed [17].

## 1.6   STRUCTURE FACTORS

### 1.6.1   Analytical Expressions

Analytical expressions for structure factors are available for a few simple systems including spherical and cylindrical particles [17, 18] and lamellar structures. For spheres, the hard sphere structure factor is the simplest model, as the name suggests this structure factor is derived based on purely steric interactions between solid packed spheres (volume fraction $\phi$ and hard sphere radius $R_{HS}$). It is written

$$S(q) = \frac{1}{1 + 24\phi G(2R_{HS}q)/(2R_{HS}q)} \tag{1.38}$$

Here

$$G(A) = \frac{\alpha(\sin A - A \cos A)}{A^2} + \frac{\beta(2A \sin A + (2 - A^2)\cos A - 2)}{A^3}$$
$$+ \frac{\gamma[-A^4\cos A + 4\{(3A^2 - 6)\cos A + (A^3 - 6A)\sin A + 6\}]}{A^5} \tag{1.39}$$

with

$$\alpha = \frac{(1 + 2\phi)^2}{(1 - \phi)^4}, \beta = \frac{-6\phi(1 + \frac{\phi}{2})^2}{(1 - \phi)^4}, \gamma = \frac{\phi\alpha}{2} \tag{1.40}$$

At $q = 0$, the Carnahan-Starling closure to the hard sphere structure factor gives the expression [11, 19]

$$S(q = 0) = \frac{(1 - \phi)^4}{(1 + 2\phi)^2 + \phi^3(\phi - 4)} \tag{1.41}$$

The sticky hard sphere potential allows for a simple attractive potential between spheres (the equation for the structure factor is presented elsewhere [17, 18]). For charged spherical particles, the screened Coulomb potential may be employed.

For cylinders, a random phase approximation (RPA, Section 1.8) equation may be used or the PRISM (polymer reference interaction site model, Section 5.8) structure factor. The RPA expression is

$$S(q) = n(\Delta\rho)^2 \frac{P_{cyl}(q)}{1 + vP_{cyl}(q)} \tag{1.42}$$

Here $n$ is the number density of cylinders, $v$ is usually treated as a fit parameter and $P_{cyl}(q)$ is the form factor of a cylinder of radius $R$ and length $L$ (see also Table 1.2):

$$P_{cyl}(q) = \int_0^{\pi/2} \left[ \frac{\sin(\frac{1}{2}qL\cos\alpha)}{\frac{1}{2}qL\cos\alpha} \cdot \frac{2J_1(qR\sin\alpha)}{qR\sin\alpha} \right]^2 \sin\alpha\, d\alpha \tag{1.43}$$

Here $J_1(x)$ denotes a first order Bessel function.

It is possible to compute $v$ from an equation for osmotic compressibility [17, 18]:

$$v = \frac{(1 + 2(B + C)^2) + 2D[1 + B + \frac{5}{4}C]}{(1 - B - C)^4} - 1 \tag{1.44}$$

Here $B = \pi R^2 Ln$, $C = 4\pi r^3 n/3$, and $D = \pi RL^2/2$, where $n$ is the number density [17].

For lamellar or smectic structures a number of structure factors have been proposed, based on the fluctuations of the layers. Figure 1.6 illustrates the thermal fluctuations that arise from the flexibility of the layers in a lamellar system. These fluctuations destroy true long-range order in all one-dimensional systems according to the Landau-Peierls instability [20].

The structure factor for a stack of $N$ fluctuating layers can be written as [21]

$$S(q) = N_{diff} + \sum_{N_k=N-2\sigma}^{N+2\sigma} x_k S_k \tag{1.45}$$

**Table 1.1**   Sequences of Bragg reflections for common structures.

| Structure | Reflections | Positional ratio |
|---|---|---|
| Lamellar | (001),(002),(003),(004),(005),(006) | $1:2:3:4:5:6$ |
| Hexagonal | (1,0),(1,1),(2,0),(2,1),(3,0),(2,2) | $1:\sqrt{3}:\sqrt{4}:$ $\sqrt{7}:\sqrt{9}:\sqrt{12}$ |
| Body-centred cubic $Im\bar{3}m$ | (110),(200),(211),(220),(310),(222) | $\sqrt{2}:\sqrt{4}:\sqrt{6}:$ $\sqrt{8}:\sqrt{10}:\sqrt{12}$ |
| Face-centred cubic $Fm\bar{3}m$ | (111),(200),(220),(311),(222),(400) | $\sqrt{3}:\sqrt{4}:\sqrt{8}:$ $\sqrt{11}:\sqrt{12}:\sqrt{16}$ |
| Bicontinuous cubic Primitive cubic 'plumber's nightmare' $Im\bar{3}m$ | As $Im\bar{3}m$ above | As $Im\bar{3}m$ above |
| Bicontinuous cubic 'double diamond' $Pn\bar{3}m$ | (110),(111),(200),(211),(220),(300) | $\sqrt{2}:\sqrt{3}:\sqrt{4}:$ $\sqrt{6}:\sqrt{8}:\sqrt{9}$ |
| Bicontinuous cubic 'gyroid' $Ia\bar{3}d$ | (211),(220),(321),(400),(420),(332),(422) | $\sqrt{6}:\sqrt{8}:\sqrt{14}:$ $\sqrt{16}:\sqrt{20}:\sqrt{24}$ |

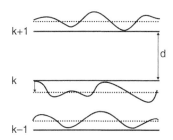

**Figure 1.6**   Fluctuations in the layer positions in a lamellar structure that are characterized by the membrane stiffness.

The term $\sigma$ is the width parameter in a Gaussian function

$$x_k = \frac{1}{\sigma\sqrt{2\pi}} \exp\left[-\frac{(N_k - N)^2}{2\sigma^2}\right] \tag{1.46}$$

This is used to account for polydispersity in the number of layers and a suitable choice is $\sigma = \sqrt{N}$ for a sufficiently large N. In Eq. (1.45), $N_{\text{diff}}$ is a diffuse scattering term arising from uncorrelated fluctuations of layers.

Different models have been proposed to describe the thermal fluctuations and hence the structure factor terms $S_k$. In the thermal disorder model, each layer fluctuates with an amplitude $\Delta = \langle (d_k - d)^2 \rangle$, where $d$ is the average lamellar spacing. The structure factor is that of an ideal one-dimensional crystal multiplied by a Debye-Waller factor:

$$S_k = N_k + 2 \exp\left(-\frac{q^2\Delta^2}{2}\right) \sum_{m=1}^{N_k-1} (N_k - m)\cos(mqd) \qquad (1.47)$$

In the second model, the paracrystalline model (this type of model is discussed further in Section 1.6.3), the position of an individual fluctuating layer in a paracrystal is determined solely by its nearest neighbours. Then [21, 22]

$$S_k = N_k + 2 \sum_{m=1}^{N_k-1} (N_k - m)\cos(mqd)\exp\left(-\frac{m^2 q^2 \Delta^2}{2}\right) \qquad (1.48)$$

In a third alternative model, introduced by Caillé [23] and modified to allow for finite lamellar stacks [24, 25], the fluctuations are quantified in terms of the flexibility of the membranes:

$$S_k = N_k + 2 \sum_{m=1}^{N_k-1} (N_k - m)\cos(mqd)\exp\left[-\left(\frac{qd}{2\pi}\right)^2 \eta\gamma\right] (\pi m)^{-(qd/2\pi)^2\eta} \qquad (1.49)$$

Here $\gamma$ is Euler's constant and

$$\eta = \pi k_B T / 2d^2 (BK)^{1/2} \qquad (1.50)$$

is the Caillé parameter, which depends on the bulk compression modulus $B$ and the bending rigidity $K$ of the layers.

## 1.6.2 Periodic Structures and Bragg Reflections

For ordered systems such as colloid crystals, liquid crystals or block copolymer mesophases, the structure factor will comprise a series of Bragg reflections (or pseudo-Bragg reflections, strictly, for layered structures). The amplitude structure factor for a $hkl$ reflection (where $h$, $k$ and $l$ are Miller indices) of a lattice is given by [12]

$$F_{hkl} = \int \rho(\mathbf{r}) \exp[-2\pi i(h\mathbf{a}^* + k\mathbf{b}^* + l\mathbf{c}^*).\mathbf{r}]d\mathbf{r} \qquad (1.51)$$

Here $\mathbf{a}^*$, $\mathbf{b}^*$, and $\mathbf{c}^*$ are reciprocal space axis vectors. In Eq. (1.51), $\rho(r)$ is used rather than $\Delta\rho(\mathbf{r})$ since if $F_{hkl}$ is determined on an absolute scale, the Fourier transform of Eq. (1.51) permits the determination of $\rho(\mathbf{r})$ in absolute units (e.g. electron $\text{Å}^{-3}$ in the case of SAXS data).

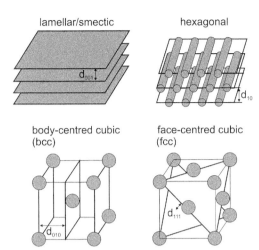

**Figure 1.7** Representative one-, two-, and three-dimensional structures, with lowest indexed diffraction planes indicated. For the lamellar structure, three-dimensional Miller indices have been employed while for the hexagonal structure two-dimensional indices are used.

The location of the observed Bragg reflections (ratio of peak positions) is characteristic of the symmetry of the structure. Table 1.1 lists the observed reflections for common structures observed for soft materials (and some hard materials). Figure 1.7 shows examples of structures with the lowest indexed planes indicated. The generating equations for allowed reflections for different space groups are available in crystallography textbooks [26] and elsewhere [27].

Figure 1.8 shows representative SAXS intensity profiles for some common ordered phases in soft materials, exemplified by data for block copolymer melts. The sequences of observed reflections are consistent with Table 1.1.

The layer spacing $d$ for a lamellar phase is given by

$$q_l = \frac{2\pi l}{d} \tag{1.52}$$

Here $q_1$ is the position of the $l$th order Bragg peak.

For a two-dimensional hexagonal structure [26, 29]:

$$q_{hk} = \frac{4\pi(h^2 + k^2 + hk)^{1/2}}{\sqrt{3}a} \tag{1.53}$$

where $q_{hk}$ is the position of the Bragg peak with indices $hk$ and $a$ is the lattice constant.

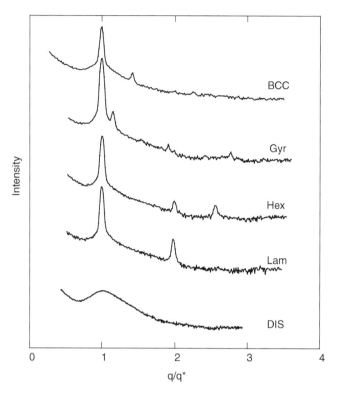

**Figure 1.8** Compilation of SAXS profiles measured for PEO-PBO [polyoxye-thylene-*b*-polyoxybutylene] diblock copolymer melts. The *x*-axis uses a *q*-scale normalized to $q^*$, the position of the first order peak. BCC: body-centred cubic, Gyr: gyroid, Hex: hexagonal-packed cylinders, Lam: lamellar, DIS: disordered. *Source*: From Ref. [28].

For a cubic structure [26, 29]

$$q_{hkl} = \frac{2\pi(h^2 + k^2 + l^2)^{1/2}}{a} \tag{1.54}$$

where $q_{hkl}$ is the position of the Bragg peak with indices *hkl* and *a* is the lattice constant.

The general expression for all crystal systems is [12, 26]

$$q_{hkl}^2 = 4\pi^2[\frac{h^2\sin^2\alpha}{a^2} + \frac{k^2\sin^2\beta}{b^2} + \frac{l^2\sin^2\gamma}{c^2} + \frac{2hk}{ab}(\cos\alpha\cos\beta - \cos\gamma)$$

$$+ \frac{2kl}{bc}(\cos\beta\cos\gamma - \cos\alpha) + \frac{2lh}{ca}(\cos\gamma\cos\alpha - \cos\beta)]/X \tag{1.55}$$

with

$$X = 1 + 2\cos\alpha\cos\beta\cos\gamma - \cos^2\alpha - \cos^2\beta - \cos^2\gamma \tag{1.56}$$

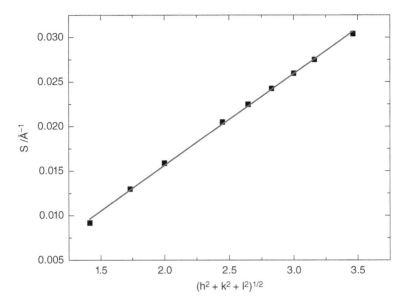

**Figure 1.9**  Indexation of SAXS reflections observed for the lipid monoolein forming a bicontinuous $Pn\overline{3}m$ cubic structure, within cubosomes. Here, $S_{hkl} = 1/d_{hkl}$ is determined from the observed position of the reflection via $S_{hkl} = q_{hkl}/2\pi$. The lattice constant $a = (97.2 \pm 1.2)$ Å is determined as the reciprocal of the gradient. *Source*: From Ref. [15].

Here $a$, $b$, $c$ are the unit cell lengths and $\alpha$, $\beta$, $\gamma$ are the unit cell angles.

Explicit expressions for $q_{hkl}$ for other lattices can be found elsewhere (see e.g. [26, 29]). Figure 1.9 shows an example of the determination of the lattice constant $a$ by use of Eq. (1.54) by indexing the observed reflections of a cubic structure (lipid bicontinuous cubic structure). Similar methods can be used to determine $d$ for lamellar structures from Eq.(1.52) or $a$ for hexagonal structures from Eq. (1.53).

## 1.6.3  Partially Ordered Systems and Paracrystals

The number of observed reflections as well as the peak width for an ordered structure gives an indication of the extent of order. Figure 1.10 illustrates this schematically for a one-dimensional lattice with variable degrees of short- or long-ranged order [20, 30].

For the case of a crystalline sample with long-range positional order, the scattering pattern will consist of a function of sharp resolution-limited diffraction peaks, as shown in Figure 1.10a. Thermal disorder leads to peak

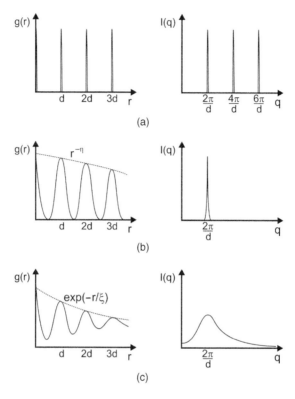

**Figure 1.10** Long-range versus short-range order showing schematics of real space distribution functions $g(r)$ (left), and scattered intensity profiles $I(q)$ (right). (a) Long-range order, (b) Quasi-long-range order, (c) Short range order.

attenuation described (in the case of isotropic disorder) by a Debye-Waller factor, $\exp\left(-\frac{q^2\langle u^2\rangle}{3}\right)$, where $\langle u^2\rangle$ is the mean-squared displacement from the scattering centre.

For a sample with quasi-long-range order, the scattering profile consists of a sharp peak with power-law tails (Figure 1.10b). More commonly observed is short-range order. This produces a scattering profile containing a Lorentzian peak (Figure 1.10c) given by

$$I(q) = \frac{I_0}{1 + \xi^2 (q - q_0)^2} \tag{1.57}$$

Here $\xi$ is a correlation length and $q_0 = 2\pi/d$ is the peak position ($d$ is the preferred spacing in the system). If the decay of the peaks in the correlation function is steeper, $g(r) \sim \exp(-r^2/\xi^2)$, then the intensity profile shows a Gaussian peak:

$$I(q) = I_0 \exp[-\xi^2 (q - q_0)^2] \tag{1.58}$$

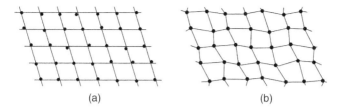

(a)                                      (b)

**Figure 1.11**   Schematic of lattice distortions in paracrystals with lattice distortions of (a) first and (b) second kind.

For more ordered or partly crystalline materials, paracrystal models can be employed in the analysis of SAS data. Paracrystals are imperfect crystals in which a degree of lattice order is retained. Figure 1.11 illustrates paracrystals of the first and second kind, for the case of a two-dimensional lattice.

In a paracrystal with imperfections of the first kind, the long-range lattice order is retained but there are fluctuations in position around the lattice nodes (Figure 1.11a), i.e. there is positional disorder. In the observed diffraction pattern, the peak width increases linearly with the peak order [31], and the intensity is modulated by a Debye-Waller factor as described above [12, 31]. With lattice distortions of the second kind, the long-range lattice order is disrupted as shown in Figure 1.11b, i.e. there is long-range positional and bond orientational disorder. This leads to a quadratic increase in peak width with peak order [12, 31]. Analytical solutions are available for one-dimensional paracrystal models of the first and second kind [12, 22].

## 1.6.4   The Phase Problem

The amplitude structure factor in Eq. (1.27) is a complex quantity and so can be written as the product of an amplitude and a phase:

$$F_{hkl} = |F_{hkl}| \exp(i\phi_{hkl}) \qquad (1.59)$$

where $\phi_{hkl}$ is the phase angle for reflection with indices $hkl$. The amplitude can be obtained from the square root of the intensity:

$$|F_{hkl}| = \sqrt{I_{hkl}} \qquad (1.60)$$

In general, the phase appearing in Eq. (1.59) cannot be determined in an x-ray diffraction or SAXS experiment. There are methods in x-ray crystallography to circumvent the problem (such as heavy atom replacement). In the study of SAXS by soft materials, the reconstruction of the electron density profile is usually only done for a limited subset of systems. In particular, it is performed for lamellar structures such as those of lipid bilayers, as discussed further in Section 4.12.

## 1.6.5   Fractal Structures

A number of structure factors have been proposed for fractal structures. For these systems, the separation of form and structure factors often does not make sense; therefore, we denote the scattering just by the intensity. One model commonly used in the analysis of SAS data is the Fischer–Burford fractal model, for which the structure factor is given by [32, 33]

$$I(q) = I_0 \left( 1 + \frac{2}{3D} q^2 R_g^2 \right)^{-D/2} \tag{1.61}$$

where $I_0$ is the forward scattering intensity, $D$ is the fractal dimension, and $R_g$ is the radius of gyration of the aggregate.

Another widely employed expression is that for the structure factor of a mass-fractal: [33–36]

$$I(q) = I_0 \frac{\sin[(D-1)\tan^{-1}(q\xi)]}{(D-1)q\xi(1+q^2\xi^2)^{(D-1)/2}} \tag{1.62}$$

where $I_0$ denotes forward scattering, $D$ is the fractal dimension, and $\xi$ is the correlation length.

Another widely used model was developed by Beaucage for a variety of systems with ordering on multiple length scales including fractal structures, and is based on an expression for the scattering from excluded volume polymer fractals [37, 38]. This gives a function that interpolates between a Guinier function at low $q$ and a Porod function at high $q$. It is used for many types of systems, such as porous materials (Section 5.11), colloidal and gel aggregate structures, polymer foams, polymer nanocomposites, nanopowders, and other systems. The unified Beaucage expression for a system with one structural level is [39]:

$$I(q) = G \exp\left( \frac{-q^2 R_g^2}{3} \right) + \frac{P}{q^D} \left[ \mathrm{erf}\left( \frac{qR_g}{6^{1/2}} \right) \right]^{3D} \tag{1.63}$$

Here $G$ is a Guinier scaling factor, $R_g$ is the radius of gyration of the mass fractal object, $P$ is a Porod constant, $D$ is the fractal dimension, and erf denotes the error function. The constant $P$ is related to $G$ via the expression [38]:

$$P = \frac{GD}{R_g^d} \left[ \frac{6D^2}{(2+D)(2+2D)} \right]^{D/2} \Gamma\left( \frac{D}{2} \right) \tag{1.64}$$

Here $\Gamma(D/2)$ denotes a Gamma function. Analytical expressions are available for $G$ and $P$ for a variety of uniform particles with different shapes as well as types of polymers [40].

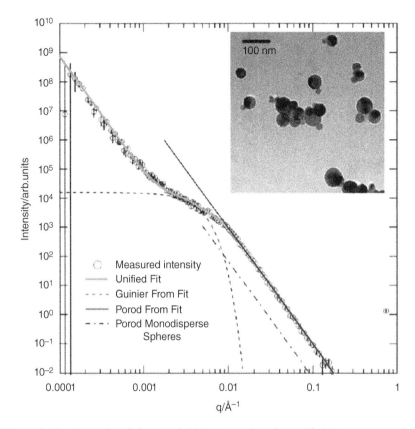

**Figure 1.12** Example of fitting of SAS data using the unified Beaucage model, Eq. (1.63), to fit USAXS data for a titania nanopowder (inset: TEM image). The contributions from the Guinier and Porod components of Eq. (1.63) are shown. Also shown for comparison is the limiting Porod slope in a scattering profile from spheres of the same radius of gyration $R_g$ as from the Beaucage model fit. *Source*: From Beaucage et al. [40]. © 2004, International Union of Crystallography.

Figure 1.12 shows an example of a fit to USAXS data for a titania nanopowder, along with indicated contributions from the Guinier and Porod components of Eq. (1.64).

A generalized Beaucage function is used for a system with two structural levels with radius of gyration $R_g$ for the fractal and $R_{sub}$ for the subunits [41]

$$I(q) = G \exp\left(\frac{-q^2 R_g^2}{3}\right) + G_s \exp\left(\frac{-q^2 R_{sub}^2}{3}\right) + \frac{P}{q^D} \exp\left(-\frac{q^2 R_s^2}{3}\right) \left[\text{erf}\left(\frac{qkR_g}{6^{1/2}}\right)\right]^{3D}$$

$$+ \frac{P_s}{q^{D_s}} \left[\text{erf}\left(\frac{qk_s R_s}{6^{1/2}}\right)\right]^{3d_s} \tag{1.65}$$

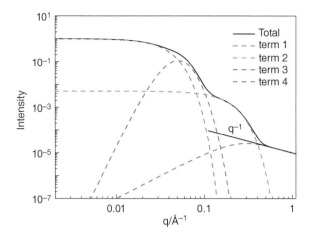

**Figure 1.13**   Example of calculated form factor using the generalized Beaucage form factor, Eq. (1.65) along with the components of this equation. Calculation for $G = 1$, $P = 1 \times 10^{-4}$, $G_s = 5 \times 10^{-3}$, $P_s = 1 \times 10^{-5}$, $R_g = 50\,\text{Å}$, $R_{sub} = 30\,\text{Å}$, $R_s = 10\,\text{Å}$, $k = 1$, $k_s = 1$, $D = 4$, $D_s = 1$ (limiting Porod slope shown). Calculated using SASfit (Table 2.2).

Here $k$ is an empirical constant, $k \approx 1$ [41], and $G_s$, $P_s$ are the Guinier and Porod scaling terms for the subunit components of the intensity, $d_s$ is the fractal dimension for the subunits with size $R_s$, and $P$ and $D$ are as defined in Eqs. (1.63) and (1.64). This function has a large number of parameters and can be used to fit SAS data from many multilevel systems. Figure 1.13 shows an example of a calculated generalized Beaucage form factor along with the components corresponding to the corresponding terms in Eq. (1.65).

The low $q$ scaling from a fractal structure shows a power law behaviour, $I(q) \sim q^{-D}$ in a range $1/R_g \ll q \ll 1/R_s$ [12]. Further details on SAS studies of fractal systems are available in a dedicated text [42].

## 1.6.6   Microemulsions

A widely used expression for the scattering from bicontinuous microemulsions is due to Teubner and Strey. The intensity is computed from a phenomenological Landau–Ginzburg free energy expansion including gradient terms of the composition order parameter. The structure factor (as in the previous section written as an intensity) is given by [43, 44]:

$$I(q) = \frac{8\pi c_2 \langle \eta^2 \rangle / \xi}{a_2 + c_1 q^2 + c_2 q^4} \tag{1.66}$$

The terms $a_2$, $c_1$, $c_2$ are coefficients in the free energy expansion and for a microemulsion, $c_1 < 0$ while $a_2 > 0$ and $c_2 > 0$. The term $\langle \eta^2 \rangle = \phi_1 \phi_2 (\Delta \rho)^2$ represents a weighted difference in scattering densities between the two phases with volume fractions $\phi_1$ and $\phi_2$. The corresponding real-space correlation function related to Eq. (1.66) by a Fourier transform is

$$\Gamma(r) = \frac{\sin(kr)}{kr} \exp\left(-\frac{r}{\xi}\right) \tag{1.67}$$

Here $k = 2\pi/d$. There are two length scales associated with the microemulsion, the correlation length given by

$$\xi = \left[\frac{1}{2}\left(\frac{a_2}{c_2}\right)^{1/2} + \frac{c_1}{4c_2}\right]^{-1/2} \tag{1.68}$$

and the domain size given by

$$d = 2\pi \left[\frac{1}{2}\left(\frac{a_2}{c_2}\right)^{1/2} - \frac{c_1}{4c_2}\right]^{-1/2} \tag{1.69}$$

Figure 1.14 shows an example of fitting SAXS data for a microemulsion using the Teubner-Strey structure factor. The intensity at high $q$ in the Teubner-Strey equation shows limiting Porod behaviour:

$$\lim_{q \to \infty} I(q) = \frac{8\pi \langle \eta^2 \rangle}{\xi} q^{-4} \tag{1.70}$$

Berk gave an alternative (more complex) expression for $\Gamma(r)$ for a random wave model, which can also be used to describe bicontinuous microemulsions [46].

For a two-phase system with a preferred correlation length, the Debye-Bueche structure factor, sometimes known as the Debye-Anderson-Brumberger model [47–49] may be used:

$$I(q) = I(0)\frac{1}{[1 + (q\xi)^2]^2} \tag{1.71}$$

In this model, the correlation function exhibits a simple exponential decay:

$$\Gamma(r) = \exp\left(-\frac{r}{\xi}\right) \tag{1.72}$$

Here again, $\xi$ is a correlation length. Comparison of Eqs. (1.72) and (1.67) shows that the Teubner-Strey equation introduces the damped periodic function $j_0(kr) = \sin(kr)/kr$ into the correlation function. Figure 1.15 shows examples of calculated intensity profiles with the Debye-Bueche

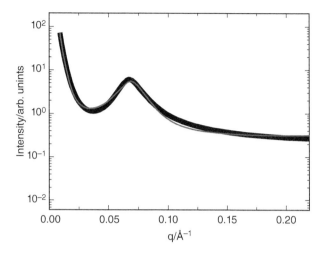

**Figure 1.14** Example of SAXS data for a bicontinuous microemulsion (Pluronic copolymer with tannic acid in aqueous solution) (open symbols), fitted to the Teubner-Strey structure factor (red line). *Source*: From Dehsorkhi et al. [45]. © 2011, Royal Society of Chemistry.

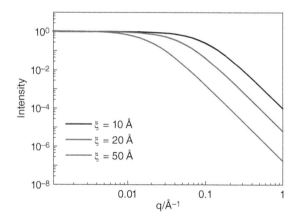

**Figure 1.15** Calculated structure factors using the Debye-Bueche structure factor with $I(0) = 0$ and the correlation lengths indicated.

structure factor. The Debye-Bueche model is obtained as a special case of the Teubner-Strey model for very large $d$-spacing ($d \gg \xi$). The intensity decays as $q^{-4}$ in the Porod regime at high $q$ [48].

Other types of microemulsion contain droplets, and the scattering can be described using models for the form and structure factors of globular objects.

## 1.6.7 Integral Equation Theories, Closures, and Structure Factors

The Ornstein-Zernike (OZ) equation is an integral equation used in liquid state theory to describe the correlation between two molecules. Specifically, the total correlation function between particles at $r_1$ and $r_2$, with inter-particle separation $r_{12}$ is written as

$$h(r_{12}) = g(r_{12}) - 1 \qquad (1.73)$$

where $g(r)$ is the radial distribution function. The OZ equation is [50]

$$h(r_{12}) = c(r_{12}) + \rho \int c(r_{13})h(r_{23})dr_3 \qquad (1.74)$$

Here $c(r)$ are direct correlation functions and $\rho$ is the density. This integral equation gives the total correlation function as the sum of the direct correlation and an integral over the indirect correlations involving a third particle, integrating over its position $r_3$.

The Fourier transform of the OZ equation is written as

$$H(q) = \frac{C(q)}{1 - \rho C(q)} \qquad (1.75)$$

Where $H(q)$ is the Fourier transform of $h(r)$ and $C(q)$ is the transform of $c(r)$.

The structure factor is related to the pair correlation function $g(r)$ via

$$S(q) = 1 + \rho \int_0^\infty [g(r) - 1]e^{iq \cdot r}dr \qquad (1.76)$$

It may be noted that this equation differs from that given in Eq. (1.32) since the spherical average is not employed, and the normalization of $g(r)$ differs.

Equation (1.76) along with Eq. (1.75) lead to the OZ expression for the structure factor:

$$S(q) = \frac{1}{1 - \rho C(q)} \qquad (1.77)$$

In the limit $\rho \to 0$ the integral in Eq. (1.74) vanishes and the direct correlation function is [50]

$$c(r) = e^{-\beta v(r)} - 1 \qquad (1.78)$$

Here $\beta = 1/k_B T$. If $\beta v(r) \ll 1$ Eq. (1.78) gives

$$c(r) = -\beta v(r) \qquad (1.79)$$

Which is known as the mean spherical approximation (MSA), used to evaluate the OZ equation.

The OZ equation can be solved using closure relationships for $c(r)$ such as the hypernetted chain (HNC) approximation and the Percus-Yevick (PY) approximation. The HNC equation is [50]

$$g(r) = \exp[-\beta v(r) + h(r) - c(r)] \qquad (1.80)$$

or

$$c(r) = \beta v(r) + h(r) - \ln[h(r) + 1] \qquad (1.81)$$

This simplifies to the MSA in the limit $r \to \infty$ since $h(r) \to 0$ in this limit, but generalized for any density.

The PY closure equation is given by

$$g(r) = e^{-\beta v(r)}[1 + h(r) - c(r)] \qquad (1.82)$$

This is obtained from an expansion of the HNC closure [50].

Exact solutions of these closure relationships are known, for instance using the PY approximation for hard spheres [50] or the Carnahan-Starling approximation gives the expression in Eq. (1.38).

Further information on integral equation theories for liquids is available elsewhere [51].

## 1.7  FORM FACTORS

### 1.7.1  Examples of Form Factor Expressions

Analytical expressions are available for the form factor of particles of many geometries, as well as polymers discussed in Section 1.8. Some examples of

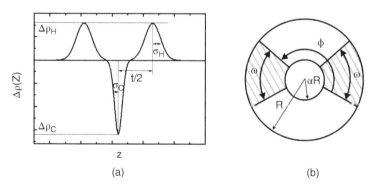

**Figure 1.16**  Parameters for complex form factors. (a) Gaussian bilayer, (b) Projection of a helical structure showing definitions of angles and radii in the corresponding Pringle–Schemidt form factor (Table 1.2).

**Table 1.2** Expressions for common form factors.

| Form factor | Equation | Parameters |
|---|---|---|
| Homogeneous sphere | $P(q) = K_s^2(q, R, \Delta\rho) = [V\Delta\rho F_s(q, R)]^2$ $F_s(q, R) = \dfrac{3[\sin(qR) - qR\cos(qR)]}{(qR)^3}$ | Radius, $R$ $\left(V = \frac{4}{3}\pi R^3\right)$, Scattering contrast, $\Delta\rho$ |
| Spherical shell | $P(q) = [K_s(q, R_1, \Delta\rho) - K_s(q, R_2, \Delta\rho(1 - \mu))]^2$ | Outer radius $R_1$ with shell scattering contrast $\Delta\rho$ Inner radius $R_2$ with core scattering contrast $\Delta\rho(1-\mu)$ |
| Gaussian polymer chain –Debye function (see Section 1.8 for derivation) | $K_s(q, R, \Delta\rho)$ as for a homogeneous sphere as above (other parameterizations are possible) $P(q) = I_0 \dfrac{2(e^{-X} + X - 1)}{X^2} = I_0 P_G(q, R_g)$ $X = (q^2 b^2 N)/6 = q^2 R_g^2$ | $I_0$, forward scattering intensity $R_g$, radius of gyration ($b$, statistical segment length) |
| Sphere with attached Gaussian chains (used for block copolymer micelles) [52, 53] | $P(q) = N_c^2(\Delta\rho_s)^2 F_s^2(q, R) + N_c(\Delta\rho_s)^2 P_G(q, R_g) +$ $N_c(N_c - 1)(\Delta\rho_c)^2 S_{cc}(q) + 2N_c^2 \Delta\rho_s \Delta\rho_c S_{sc}(q)$ where $S_{cc}(q) = \psi^2(q, R_g) \left[ \dfrac{\sin[q(R + dR_g)]}{q(R + dR_g)} \right]^2$ and $S_{sc}(q) = F_s(q, R)\psi(q, R_g) \dfrac{\sin[q(R + dR_g)]}{q(R + dR_g)}$ with $\psi(q, R_g) = \dfrac{1 - e^{-X}}{X}$ $X = (q^2 b^2 N)/6 = q^2 R_g^2$ | $N_c$, aggregation number $R$, core radius $R_g$, radius of gyration of attached chains $d$, displacement of chains (for non-penetration into core, $d \approx 1$) $\Delta\rho_s$, scattering contrast of spherical core $\Delta\rho_c$, scattering contrast of attached chains |

(continued)

**Table 1.2** *(continued)*

| Form factor | Equation | Parameters |
|---|---|---|
| Ellipsoid of revolution (spheroid) | $P(q) = \int [K_s(q, \sqrt{R_p^2 y^2 + \nu^2 R_p^2 (1+y)^2}, \Delta\rho)]^2 \, dy$ <br><br> $K_s(q, R, \Delta\rho)$ as for a homogeneous sphere as above (other parameterizations are possible) | Radius in polar direction $R_p$ <br> Equatorial radius $R_e = \nu R_p$ <br><br> Scattering contrast, $\Delta\rho$ |
| Tri-axial ellipsoid | $P(q) = \int [K_s(q, R(a, b, c, \alpha, \beta, \Delta\rho)]^2 \sin\alpha d\alpha d\beta$ <br><br> $R(a, b, c, \alpha, \beta)$ <br> $= [(a^2 \sin^2\beta + b^2 \cos^2\beta)\sin^2\alpha + c^2 \cos^2\alpha]^{1/2}$ <br><br> $K_s(q, R, \Delta\rho)$ as for a homogeneous sphere as above (other parameterizations are possible) | Semi axis lengths $a$, $b$, $c$ <br><br> Scattering contrast, $\Delta\rho$ |
| Rectangular parallelepipedons (including cubes) | $P(q) =$ <br> $V^2 (\Delta\rho)^2 \int_0^{\pi/2} \int_0^{\pi/2} [j_0(qa')i_0(qb')i_0(qc')]^2 \sin\alpha d\alpha d\beta$ <br><br> Where $j_0(qa') = \dfrac{\sin(qa \sin\alpha \cos\beta)}{qa \sin\alpha \cos\beta}$ <br><br> $i_0(qb') = \dfrac{\sin(qb \sin\alpha \sin\beta)}{qb \sin\alpha \sin\beta}$ <br><br> $i_0(qc') = \dfrac{\sin(qc \cos\alpha)}{qc \cos\alpha}$ | Edge lengths $a$, $b$, $c$ <br><br> Scattering contrast, $\Delta\rho$ |
| Circular cylinder | $P(q) =$ <br><br> $V^2 (\Delta\rho)^2 \int_0^{\pi/2} \left[ \dfrac{\sin\left(\frac{1}{2}qL\cos\alpha\right)}{\frac{1}{2}qL\cos\alpha} \cdot \dfrac{2J_1(qR\sin\alpha)}{qR\sin\alpha} \right]^2 \sin\alpha d\alpha$ | Radius, $R$. length, $L$. <br> Scattering contrast, $\Delta\rho$ |

Infinitely thin circular disc

$$P(q) = V^2 (\Delta\rho)^2 \frac{2}{q^2 R^2} \left[ 1 - \frac{J_1(2qR)}{qR} \right] = P_{disc}(q)$$

Radius, $R$
Scattering contrast, $\Delta\rho$

Lipid bilayer with
Gaussian scattering density profile
(within a disc)

$$P(q) = [P_{disc}(q)/(\Delta\rho)^2]|P_{cs}(q)$$

where the cross-section form factor $P_{cs}(q)$ is

Cross-section parameters
illustrated in Figure 1.16

$$P_{cs}(q) = \left[ 2\sqrt{2\pi}\sigma_H \Delta\rho_H \exp\left( -\frac{(q\sigma_H)^2}{2} \right) \cos\left( \frac{qt}{2} \right) \right.$$
$$\left. + \sqrt{2\pi}\sigma_C \Delta\rho_C \exp\left( -\frac{(q\sigma_C)^2}{2} \right) \right]^2$$

Helices and helical tapes
(Pringle–Schmidt form factor [54])

$$I(q) =$$

$$\sum_{n=0}^{\infty} \varepsilon_n \cos^2\left( \frac{n\phi}{2} \right) \frac{\sin^2\left( \frac{n\omega}{2} \right)}{(n\omega/2)^2} \frac{1}{2} \int_0^\pi |G_n(q,\alpha)|^2 \sin\alpha\, d\alpha$$

$H$ helix length, $P$ helix pitch,
other parameters defined in
Figure 1.16

Here

$$G_n(q,\alpha) = \frac{\int_{\alpha R}^R J_n(qr\sin\alpha) \dfrac{\sin\left[ \dfrac{H}{2}\left( q\cos\alpha + \dfrac{2\pi n}{P} \right) \right]}{\dfrac{H}{2}\left( q\cos\alpha + \dfrac{2\pi n}{P} \right)} r\, dr}{\frac{1}{2}R^2(1-\alpha^2)}$$

and $\varepsilon_0 = 1$ and $\varepsilon_n = 2$ for $n \geq 1$

more commonly used form factors are listed in Table 1.2. Extensive compilations of form factors are available [7, 17, 18, 21, 55].

The form factors for particulate systems of different dimensionality can all be expressed in terms of hypergeometric functions [27].

## 1.7.2   Limiting Behaviours

The scaling behaviour of the intensity (plotted on a double logarithmic scale) at low $q$ for monodisperse and uniform spheres, long cylinders and flat particles (disc or layer structures) is shown in Figure 1.17, along with examples of calculated form factors (examples of experimental data corresponding to this type of structure are presented in Section 4.13). For spherical particles, the slope is almost zero (it cannot be exactly zero according to the Guinier equation, Eq. (1.24)). For a long cylinder $I \sim q^{-1}$ at low $q$ whereas for a flat particle such as a disc or a bilayer structure $I \sim q^{-2}$ at low $q$.

These scaling behaviours for extended rod-like and flat particles can be derived as discussed in the following section.

Alternatively, the scaling behaviour can be obtained from the behaviours of the autocorrelation functions at large $r$ (relating to low $q$ behaviour of the intensity). For cylinders (radius $R$), [56]

$$\gamma(r) \approx \left( \frac{2\pi R^2}{4\pi r^2} \right) = \frac{1}{2} \left( \frac{R}{r} \right)^2 \tag{1.83}$$

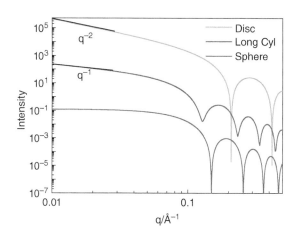

**Figure 1.17**   Form factors calculated for homogeneous particles along with limiting slopes. Form factors are calculated for spheres of radius $R = 30\,\text{Å}$, cylinders of radius $R = 30\,\text{Å}$ and length $L = 1000\,\text{Å}$ and discs with thickness $T = 30\,\text{Å}$ and radius $R = 1000\,\text{Å}$. The calculated profiles have been offset vertically for convenience. The minima in principle have zero intensity, but are truncated due to numerical calculation accuracy and for convenience plotting on a logarithmic intensity scale. The profiles were calculated using SASfit [21].

For flat particles (thickness $T$) [56]

$$\gamma(r) \approx \left(\frac{2\pi r T}{4\pi r^2}\right) = \frac{T}{2r} \tag{1.84}$$

Substitution of these expressions into Eq. (1.8) leads to the following equations: [56]

$$I(q) = \frac{\pi^2(\Delta\rho)^2 R^2}{q} \sim q^{-1} \text{ for cylinders} \tag{1.85}$$

$$I(q) = \frac{2\pi(\Delta\rho)^2 T}{q^2} \sim q^{-2} \text{ for discs/layers} \tag{1.86}$$

For monodisperse spherical particles of radius $R$, the minima in the form factor are located at $qR = 4.493, 7.725...$ [10]. For an isotropic system of long cylindrical rods (radius $R$), the form factor minima are located at $qR = 3.83, 7.01 ...$ [10] and for flat particles (thickness $T$) at $qT/2 = 3.14$, $6.28 ...$ These values can be obtained from the minima in the corresponding Bessel or sine functions as in the equations in Section 1.7.3.

It should be noted that the form factors and the position of the minima in them depend on products of $q$ and an appropriate particle dimension; therefore, the form factors have the same appearance for different pairs of reciprocal units (i.e. calculations with $q$ in Å$^{-1}$ with dimensions in Å or $q$ in nm$^{-1}$ with dimensions in nm give the same result).

## 1.7.3 Factoring Scattering from the Particle Cross-Section

For a long cylindrical particle the scattering intensity (form factor) can be computed from Eq. (1.6) by calculating the average $\exp[-i\mathbf{q}.\mathbf{r}]$ in polar coordinates $(l, z, \gamma)$, where the $\mathbf{q}$ vector and vector $\mathbf{r}$ to a point in the cylinder are related by the polar angles $(\phi, \gamma)$ (Figure 1.18):

$$\langle \exp[-i\mathbf{q}.\mathbf{r}] \rangle = \langle \exp[-iq(z\cos\phi - l\sin\phi\cos\gamma)] \rangle \tag{1.87}$$

For a uniform cylinder (radius $R$, length $L$) this leads to the expression [57]

$$I(q) = (\Delta\rho)^2 \left\langle \left( \int_{z=-\frac{L}{2}}^{z=\frac{L}{2}} \int_{l=0}^{l=R} \int_{\gamma=0}^{\gamma=2\pi} \exp[-iq(z\cos\phi - l\sin\phi\cos\gamma)]d\gamma ldl dz \right)^2 \right\rangle_\phi \tag{1.88}$$

$$= (\Delta\rho)^2 \left\langle \left( \int_{z=-\frac{L}{2}}^{z=\frac{L}{2}} \exp[-iqz\cos\phi]dz \int_{l=0}^{l=R} \int_{\gamma=0}^{\gamma=2\pi} \exp[-iql\sin\phi\cos\gamma]d\gamma ldl \right)^2 \right\rangle_\phi \tag{1.89}$$

Performing the integral over $z$ (using the same formula as in Eq. (1.12)) we have

$$I(q) = (\Delta\rho)^2 \left\langle \left( \frac{\sin\left(\frac{1}{2}qL\cos\phi\right)}{\frac{1}{2}qL\cos\phi} \int_{l=0}^{l=R} \int_{\gamma=0}^{\gamma=2\pi} \exp[-iql\sin\phi\cos\gamma]d\gamma ldl \right)^2 \right\rangle_\phi \tag{1.90}$$

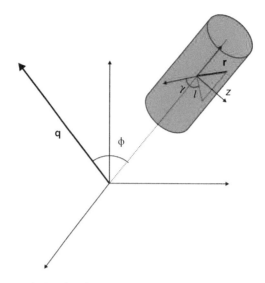

**Figure 1.18** Inter-relationship between vectors **q** and **r** in cylindrical co-ordinates.

For $L >> R$ we can factor this as the product of intensity associated with the length of the particle $I_L(q)$ and that of the cross-section $I_c(q)$ (i.e. the intensity is written as a convolution product):

$$I(q) = (\Delta\rho)^2 I_L(q) I_c(q) \tag{1.91}$$

where

$$I_L(q) = \left\langle \left| \frac{\sin\left(\frac{1}{2}qL\cos\phi\right)}{\frac{1}{2}qL\cos\phi} \right|^2 \right\rangle_\phi = \int_{\phi=0}^{\phi=\pi} \left( \frac{\sin\left(\frac{1}{2}qL\cos\phi\right)}{\frac{1}{2}qL\cos\phi} \right)^2 \sin\phi\, d\phi$$

$$\simeq \int_{x=0}^{x=\infty} \left( \frac{\sin\left(\frac{1}{2}qLx\right)}{\frac{1}{2}qLx} \right)^2 dx = \frac{L\pi}{q} \tag{1.92}$$

The integral over $\phi$ extends to infinity to make use of the Dirichlet integral, this is valid when $q \geq 2\pi/L$, since the integrand is negligibly small for $x > 1$ [7]. This leads to the factorization [5, 6]

$$I(q) = \frac{L\pi}{q} I_c(q) \tag{1.93}$$

where $I_c(q)$ is the form factor of the cross-section. As mentioned above, this is valid when $L >> R$. Eq. (1.93) shows the $I(q) \sim q^{-1}$ scaling for a cylindrical particle at low $q$ (where $I_L(q)$ dominates). This derivation is for a uniform cylinder for which $I_c(q)$ can be evaluated as in the following, however Eq. (1.93) applies in general for other rod-like particles.

The cross-section intensity is related to the distance distribution function of the cross-section, $\gamma_c(r)$, via the expression [6, 7]

$$I_c(q) = 2\pi \int_0^D \gamma_c(r) J_0(qr) dr \qquad (1.94)$$

Here $D$ is the cross-section diameter. For a uniform cylinder this may be evaluated to give [6]

$$I_c(q) = (2\pi R)^2 \left( \frac{J_1(qR)}{qR} \right)^2 \qquad (1.95)$$

In these equations $J_0(qR)$ and $J_1(qR)$ denote Bessel functions of integral order. The cross-section radius is given by $R_c = R/\sqrt{2}$ [58] (see also the discussion in Section 1.4).

The pair distribution function of the cross-section can be obtained from the cross-section intensity via an inverse Hankel transform [7]

$$\gamma_c(r) = \frac{1}{2\pi} \int_0^\infty I_c(q) J_0(qr) q dq \qquad (1.96)$$

For flat particles (discs of area $A$) the intensity can be factored, via an equation analogous to Eq. (1.93) as [4, 6, 10]

$$I(q) = \frac{2\pi A}{q^2} I_t(q) \qquad (1.97)$$

where $I_t(q)$ is the cross-section scattering that depends on the thickness $T$, which is related to the cross-section radius via $R_c = T/\sqrt{12}$ [58].

Expressions (1.93) and (1.97) indicate that as an alternative to the Guinier equation which provides $R_g$, the cylinder cross-section radius can be obtained from a plot of $\ln[qI(q)]$ vs. $q^2$ and for discs/planar structures $R_c$ can be obtained a plot of $\ln[q^2 I(q)]$ vs. $q^2$ [58, 59].

If the intensity is measured on an absolute scale it is possible to determine the mass per unit length for rod-like particles or the mass per unit area for flat (lamellar) particles. The general expression for molar mass determination for an arbitrary particle from the differential scattering cross-section $d\sigma/d\Omega$ in absolute units (cm$^{-1}$) is discussed in Section 2.9. For rod-like particles in a solution of concentration $c$ (in g cm$^{-3}$), the mass per unit length $M_c$ (in g mol$^{-1}$ cm$^{-1}$) can be obtained from the expression (containing a $\pi/q$ factor from Eq. (1.95))

$$M_c = \left[\frac{d\sigma}{d\Omega} \cdot \frac{q}{\pi}\right]_{q\to0} \cdot \frac{N_A}{c(\Delta\rho v_p)^2} \tag{1.98}$$

where $N_A$ is Avogadro's number, $v_p$ is the specific volume in cm$^3$ g$^{-1}$, $\Delta\rho$ is the contrast in cm$^{-2}$ and $q$ is in cm$^{-1}$.

For flat particles, the area per unit length $M_t$ (in g mol$^{-1}$ cm$^{-2}$) is obtained from the analogous equation (with $q^2$ dependence cf. Eq. (1.93))

$$M_t = \left[\frac{d\sigma}{d\Omega} \cdot \frac{q^2}{2\pi}\right]_{q\to0} \cdot \frac{N_A}{c(\Delta\rho v_p)^2} \tag{1.99}$$

The derivations of these equations can be found elsewhere [60].

## 1.7.4   Effect of Polydispersity

Particle polydispersity has a considerable influence on the shape of the form factor. It is included via integration over a particle size distribution $D(R')$:

$$I(q) = \int_0^\infty D(R')P_0(q, R')dR' \tag{1.100}$$

Here the subscript 0 has been added to the form factor, in order to emphasize that this is the term for the monodisperse system.

The effect of polydispersity for the example of the form factor of a uniform sphere of radius $R$ is illustrated in Figure 1.19. The polydispersity in this case is represented by a Gaussian function:

$$D(R') = c.\exp\left(-\frac{(R' - R)^2}{2\sigma^2}\right) \tag{1.101}$$

Here $\sigma$ is the standard deviation, which is related to the full width at half maximum by FWHM $= 2\sqrt{2\ln 2}\sigma$, and $c$ is a normalization constant. Other forms of distribution can be used and may be motivated by known dispersity distributions (based on the system synthesis, for example), for example log-normal functions, Schulz-Zimm functions etc.

It is clear from the example of calculated form factors for a uniform sphere in Figure 1.19 that increasing $\sigma$ causes the form factor oscillations to get progressively washed out such that they are largely eliminated for $\sigma = 7.5$ Å (25% of the radius $R$).

It can be difficult in some cases (in particular, when measurements extend over only a small $q$ range and/or for highly polydisperse systems) to disentangle polydispersity from the effect of particle anisotropy, which can also smear

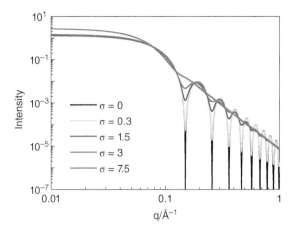

**Figure 1.19** Influence of polydispersity on the form factor of a sphere with $R = 30\,\text{Å}$ in terms of a Gaussian standard deviation with width $\sigma$ (in Å) indicated.

out form factor minima. This is illustrated by the example of Figure 1.20 which compares the calculated form factor for a sphere ($R = 30\,\text{Å}$, $\sigma = 1.5$) with that of a monodisperse ellipsoid with $R_1 = 29\,\text{Å}$ and $R_2 = 34.8\,\text{Å}$. The form factors are quite similar at low $q$ (and this example has not been optimized for maximal similarity). Especially with lower resolution SANS experiments (see Section 5.5.1 for a discussion of the resolution function for SANS), the two profiles in Figure 1.20 are not distinguishable at low $q$.

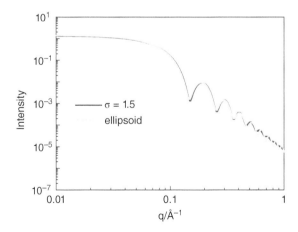

**Figure 1.20** Comparison of form factor of a polydisperse sphere ($R = 30\,\text{Å}$, $\sigma = 1.5$) and a monodisperse ellipsoid with $R_1 = 29\,\text{Å}$ and $R_2 = 34.8\,\text{Å}$.

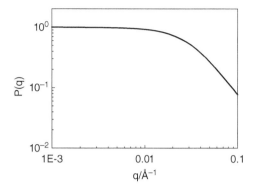

**Figure 1.21**   Debye form factor for a polymer with $R_g = 50\,\text{Å}$.

# 1.8   FORM AND STRUCTURE FACTORS FOR POLYMERS

Polymers adopt coiled conformations. In the simplest picture, these are described as ideal Gaussian coils. Polymers in the melt or in solution (under theta conditions) adopt this conformation which is the most basic model of polymer conformation. The form factor for Gaussian coils can easily be calculated (as follows) to yield the Debye function. By analogy with Eq. (1.18), but integrating over a Gaussian distribution the intensity (form factor) is [61]

$$P(q) = \frac{1}{N^2} \sum_i \sum_j \int_0^\infty \frac{\sin(qR_{ij})}{qR_{ij}} p_{Gauss}(|i-j|, R_{ij}) 4\pi R_{ij}^2 dR_{ij} \qquad (1.102)$$

The Gaussian function for a random coil depends on the end-to-end distance between points $i$ and $j$, $R_{ij}$ and $|i-j|$, which is the number of links between these points:

$$p_{Gauss}(|i-j|, R_{ij}) = \left( \frac{3}{2\pi |i-j| b^2} \right)^{3/2} \exp\left( -\frac{3R_{ij}^2}{2|i-j| b^2} \right) \qquad (1.103)$$

The integral over $R_{ij}$ can be evaluated using

$$\int_0^\infty R_{ij} \sin(qR_{ij}) \exp\left( -\frac{R_{ij}^2}{x} \right) dR_{ij} = \frac{\pi^{1/2} q x^{3/2}}{4} \exp\left( -\frac{q^2 x}{4} \right) \qquad (1.104)$$

where $x = (2|i-j|b^2)/3$. Substituting in Eq. (1.102) we obtain

$$P(q) = \frac{1}{N^2} \sum_i \sum_j \exp\left( -\frac{q^2 b^2 |i-j|}{6} \right) \qquad (1.105)$$

For a large chain ($N$ large), the summations can be replaced by integrals

$$P(q) = \frac{1}{N^2} \int_0^N \int_0^N \exp\left(-\frac{q^2 b^2}{6} \mid u - v \mid\right) du\, dv \qquad (1.106)$$

This can be evaluated [61] to give

$$P(q) = \frac{2(e^{-X} + X - 1)}{X^2} \qquad (1.107)$$

Here $X = (q^2 b^2 N)/6 = q^2 R_g^2$ ($b$ denotes a statistical segment length). Eq. (1.107) is the Debye form factor function for a polymer. Figure (1.21) shows an example of a computed Debye function. This function has a Guinier-like asymptote at low $q$, but at high $q$ ($qR_g \gg 1$), $P(q) = 2/(q^2 R_g^2)$, i.e. it shows a scaling:

$$P(q) \sim q^{-2} \qquad (1.108)$$

The expressions in Eqs. (1.107) (1.108) are valid for dilute polymer solutions under theta (ideal solution) conditions. However, this form factor and scaling behaviour is most commonly observed for polymers in the melt, which can adopt an ideal conformation since the local density within a polymer in the melt is the same as that at the macroscopic scale meaning that individual polymer chains are not perturbed by inter-chain interactions [20, 62]

The Debye form factor can be generalized to the case of expanded/collapsed coils for which $R_g = bN^\nu$, where $\nu$ is the Flory exponent ($\nu = 1/2$ for Gaussian coils such as polymer chains in a theta solvent, $\nu \approx 3/5$ for polymers in good solvents, $\nu = 1/3$ in a poor solvent). For a polymer with excluded volume, the intensity scales approximately as

$$I(q) \sim q^{-5/3} \qquad (1.109)$$

A closed-form expression for the form factor is not available in this case, although approximate equations have been presented [63]. A more accurate exponent for polymers in good solvents is $\nu = 0.588$ rather than $\nu = 0.6$, this being obtained from sophisticated theoretical analysis (renormalization group theory) [61]. Figure 1.22 compares the overall form factors of random coils (Gaussian chains) and those for excluded volume chains, indicating the Guinier regime, the power law scaling regimes and the high $q$ regime where a scaling $I(q) \sim q^{-1}$ from local stiffness is observed, this being the scaling behaviour for a rod (Eq.(1.85)(1.77)).

For polymers in solution with polymer volume fraction $\phi$, or binary blends of polymers with volume fraction $\phi$ of chain A with degree of polymerization $N_A$ (and $1 - \phi$, $N_B$ for chain B and Flory-Huggins interaction parameter of

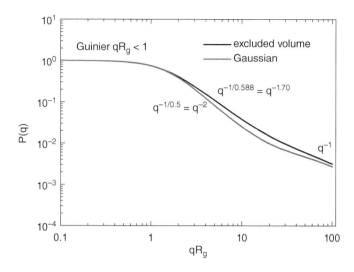

**Figure 1.22** Comparison of form factor for random walk (Gaussian) chains and excluded volume chains with scaling laws indicated, also showing Guinier regime and effect of local stiffness ($q^{-1}$ scaling of intensity) at very high $q$.

the pair, $\chi$), the structure factor at $q = 0$ can be obtained from the general compressibility relationship in terms of the free energy change on mixing $\Delta F_{mix}$:

$$S(0) = k_B T \left( \frac{\partial^2 \Delta F_{mix}}{\partial \phi^2} \right)^{-1} \tag{1.110}$$

This is related to Eq. (2.8), first noting that the dimensionless structure factor (note $q$, $V$, and $\Delta\rho$ should be in consistent units) at $q = 0$ is given by

$$S(0) = \frac{1}{V(\Delta\rho)^2} \left( \frac{d\sigma}{d\Omega} \right)_{q=0} \tag{1.111}$$

where $V$ is the volume of the system and using the relationship between isothermal compressibility $\chi_T$ and osmotic pressure, $\pi$ [12]:

$$\chi_T = - \left( \frac{1}{V} \right) \left( \frac{\partial V}{\partial \pi} \right)_T \tag{1.112}$$

The volume fraction can be written [12, 49]

$$\phi = \frac{N v_m}{V} \tag{1.113}$$

where $N/V$ is the number density and $v_m$ molar volume (of solute in the case of a polymer solution). Use of Eq. (2.8) (with $\Delta\rho$ for the contrast in a solution) leads to [49]

$$S(0) = k_B T \phi \left( \frac{\partial \pi}{\partial \phi} \right)_T^{-1} \tag{1.114}$$

Using the relationships between osmotic pressure and chemical potential and recalling that chemical potential can be written as a free energy gradient, Eq. (1.114) may be obtained from Eq. (1.112). The details of the derivation are available elsewhere [12, 49].

Using the free energy change of mixing $\Delta F_{mix}$ from Flory-Huggins theory gives

$$\frac{1}{S(0)} = \frac{1}{\phi N_A} + \frac{1}{(1-\phi)N_B} - 2\chi \tag{1.115}$$

This is generalized for non-zero wavenumbers using the random-phase approximation (RPA) to give [12, 49, 61, 64]

$$\frac{1}{S(q)} = \frac{1}{\phi N_A P(q, N_A)} + \frac{1}{(1-\phi)N_B P(q, N_B)} - 2\chi \tag{1.116}$$

Here $P(q, N)$ is the corresponding form factor of an ideal chain, i.e. the Debye function (Eq. (1.107)). These expressions apply to the case of polymers in solution, the subscripts A and B referring to solvent and solute. These expressions are also useful for the case of blends of protonated and deuterated polymers in SANS studies (see Section 5.8), where A and B label the respective unlabelled and labelled chains. The RPA is a mean field method, widely employed in polymer science to calculate the structure factor in terms of the form factor of single chains [62].

SAS data from blends is often fitted using the Ornstein-Zernike function:

$$S(q) = \frac{S(0)}{1 + (q\xi)^2} \tag{1.117}$$

This expression can be derived from Eq. (1.116) in the limit of small $q$, [62, 65] with

$$S(0) = \frac{1}{\phi N_A + (1-\phi)N_B - 2\chi} \tag{1.118}$$

The correlation length in Eq. (1.117) can be shown [61] to be

$$\xi = \sqrt{\frac{b^2 S(0)}{12\phi(1-\phi)}} \tag{1.119}$$

Analogous equations to Eq. (1.116) have been obtained for block copolymer melts, also using the random-phase approximation. The result for a diblock copolymer (degree of polymerization N, Flory-Huggins interaction parameter $\chi$ and volume fraction of one component $f$) is [28, 66]

$$S(q) = \frac{N}{F(X) - 2\chi N} \tag{1.120}$$

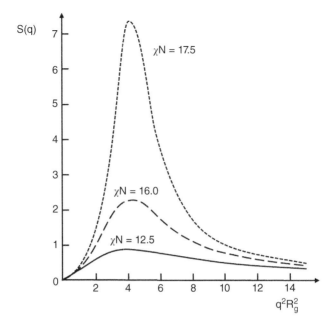

**Figure 1.23** Structure factor for a diblock copolymer melt with $f = 0.25$ at three values of $\chi N$ indicated [66]. The order-disorder transition within this model occurs at $\chi N = 17.6$ at this composition. *Source:* From Leibler [66]. © 1980, American Chemical Society.

Here $F(X)$ is a combination of Debye functions (cf. Eq. (1.107), $X$ is defined after this equation) as follows:

$$g(f, X) = \frac{2(e^{-fX} + fX - 1)}{X^2} \tag{1.121}$$

and

$$F(X) = \frac{g(1, X)}{g(f, X)g(1 - f, X) - \frac{1}{4}[g(1, x) - g(f, X) - g(1 - f, X)]^2} \tag{1.122}$$

Figure 1.23 shows an example of structure factors for a diblock copolymer in a disordered melt at several $\chi N$ values, calculated using Eq. (1.120). Since $\chi$ is inversely proportional to temperature, the $S(q)$ functions become less intense and broader as temperature increases, as expected.

Further information about scattering from block copolymers can be found elsewhere [28, 67, 68].

# REFERENCES

1. van de Hulst, H.C. (1957). *Light Scattering by Small Particles*. New York: Dover.
2. Brown, W. (ed.) (1996). *Light Scattering - Principles and Development*. Oxford: Oxford University Press.
3. Lindner, P. and Zemb, T. (eds.) (2002). *Neutrons, X-Rays and Light. Scattering Methods Applied to Soft Condensed Matter*. Amsterdam: Elsevier.
4. Glatter, O. (2018). *Scattering Methods and their Application in Colloid and Interface Science*. Amsterdam: Elsevier.
5. Guinier, A. and Fournet, G. (1955). *Small Angle Scattering of X-Rays*. New York: Wiley.
6. Glatter, O. and Kratky, O. (eds.) (1982). *Small Angle X-Ray Scattering*. London: Academic.
7. Feigin, L.A. and Svergun, D.I. (1987). *Structure Analysis by Small-Angle X-Ray and Neutron Scattering*. New York: Plenum.
8. Hamley, I.W. (1991). Scattering from uniform, cylindrically symmetric particles in liquid crystal phases. *The Journal of Chemical Physics* 95: 9376–9383.
9. Svergun, D.I. and Koch, M.H.J. (2003). Small-angle scattering studies of biological macromolecules in solution. *Reports on Progress in Physics* 66: 1735–1782.
10. Glatter, O. (2002). The inverse scattering problem in small-angle scattering. In: *Neutrons, X-Rays and Light Scattering Methods Applied to Soft Condensed Matter* (eds. P. Lindner and T. Zemb), 73–102. Amsterdam: Elsevier.
11. Svergun, D.I., Koch, M. H. J, Timmins, P.A., and May, R.P. (2013). *Small Angle X-ray and Neutron Scattering from Solutions of Biological Macromolecules*. Oxford: Oxford University Press.
12. Roe, R.-J. (2000). *Methods of X-Ray and Neutron Scattering in Polymer Science*. New York: Oxford University Press.
13. Glatter, O. (2018). *Scattering Methods and their Application in Colloid and Interface Science, Chap.3*. Amsterdam: Elsevier.
14. Brennich, M., Pernot, P., and Round, A. (2017). How to Analyze and Present SAS Data for Publication. In: *Biological Small Angle Scattering: Techniques, Strategies and Tips*, Advances in Experimental Medicine and Biology (ed. B. Chaudhuri), 47–64. Berlin: Springer.
15. Castelletto, V., Edwards-Gayle, C.J.C., Hamley, I.W. et al. (2019). Self-assembly of a catalytically active lipopeptide and its incorporation into cubosomes. *ACS Applied Bio Materials* 2: 3639–3648.
16. Kotlarchyk, M. and Chen, S.-H. (1983). Analysis of small-angle scattering spectra from polydisperse interacting colloids. *The Journal of Chemical Physics* 79: 2461–2469.
17. Pedersen, J.S. (2002). Modelling of small-angle scattering data. In: *Neutrons, X-Rays and Light Scattering Methods Applied to Soft Condensed Matter* (eds. P. Lindner and T. Zemb). Amsterdam: Elsevier.
18. Pedersen, J.S. (1997). Analysis of small-angle scattering data from colloids and polymer solutions: modeling and least-squares fitting. *Advances in Colloid and Interface Science* 70: 171–210.
19. Foffi, G., Savin, G., Bucciarelli, S. et al. (2014). Hard sphere-like glass transition in eye lens alpha-crystalline solutions. *Proceedings of National Academy of Sciences of the United States of America* 111: 16748–16753.
20. Hamley, I.W. (2007). *Introduction to Soft Matter*. Revised Edition. Chichester: Wiley.
21. Bressler, I., Kohlbrecher, J., and Thünemann, A.F. (2015). SASfit: a tool for small-angle scattering data analysis using a library of analytical expressions. *Journal of Applied Crystallography* 48: 1587–1598.

22. Hosemann, R. and Bagchi, S.N. (1962). *Direct Analysis of Diffraction by Matter*. Amsterdam: North-Holland.
23. Caillé, A. X-ray scattering by smectic-A crystals. *Comptes Rendus Hebdomaires des Seances de l'Academie des Sciences, Series B* 274: 891–892.
24. Zhang, R.T., Suter, R.M., and Nagle, J.F. (1994). Theory of the structure factor of lipid bilayers. *Physical Review E* 50: 5047–5060.
25. Pabst, G., Rappolt, M., Amenitsch, H., and Laggner, P. (2000). Structural information from multilamellar liposomes at full hydration: full q-range fitting with high quality x-ray data. *Physical Review E* 62: 4000–4009.
26. Giacovazzo, C., Monaco, H.L., Viterbo, D. et al. (1992). *Fundamentals of Crystallography*. Oxford: International Union of Crystallography/Oxford University Press.
27. Förster, S., Timmann, A., Konrad, M. et al. (2005). Scattering functions of ordered mesoscopic materials. *Journal of Physical Chemistry B* 109: 1347–1360.
28. Hamley, I.W. and Castelletto, V. (2004). Small-angle scattering of block copolymers in the melt, solution and crystal states. *Progress in Polymer Science* 29: 909–948.
29. Cullity, B.D. and Stock, S.R. (2001). *Elements of X-Ray Diffraction*. Prentice Hall: Upper Saddle River, New Jersey.
30. de Jeu, W.H. (2016). *Basic X-Ray Scattering for Soft Matter*. Oxford: Oxford University Press.
31. Stribeck, N. (2007). *X-Ray Scattering of Soft Matter*. Berlin: Springer-Verlag.
32. Fisher, M.E. and Burford, R.J. (1967). Theory of critical-point scattering and correlations I. The Ising Model. *Physics Review* 126: 583–622.
33. Sorensen, C.M. and Wang, G.M. (1999). Size distribution effect on the power law regime of the structure factor of fractal aggregates. *Physical Review E* 60: 7143–7148.
34. Sinha, S.K., Freltoft, T., and Kjems, J. (1984). Observation of power-law correlations in silica-particle aggregates by small-angle neutron scattering. In: *Kinetics of Aggregation and Gelation* (eds. F. Family and D.P. Landau). Amsterdam: Elsevier.
35. Mildner, D.F.R. and Hall, P.L. (1986). Small-angle scattering from porous solids with fractal geometry. *Journal of Physics D: Applied Physics* 19: 1535–1545.
36. Freltoft, T., Kjems, J.K., and Sinha, S.K. (1986). Power-law correlations and finite-size effects in silica particle aggregates studied by small-angle neutron-scattering. *Physical Review B* 33: 269–275.
37. Benoit, H. (1957). La diffusion de la lumière par des macromolécules en chaînes en solution dans un bon solvant. *Comptes Rendus Hebdomadaires des Seances de l'Academie des Sciences* 245: 2244–2247.
38. Hammouda, B. (2010). Analysis of the Beaucage model. *Journal of Applied Crystallography* 43: 1474–1478.
39. Beaucage, G. (1995). Approximations leading to a unified exponential power-law approach to small-angle scattering. *Journal of Applied Crystallography* 28: 717–728.
40. Beaucage, G., Kammler, H.K., and Pratsinis, S.E. (2004). Particle size distributions from small-angle scattering using global scattering functions. *Journal of Applied Crystallography* 37: 523–535.
41. Beaucage, G. (1996). Small-angle scattering from polymeric mass fractals of arbitrary mass-fractal dimension. *Journal of Applied Crystallography* 29: 134–146.
42. Anitas, E.M. (2019). *Small-Angle Scattering (Neutrons, X-Rays, Light) from Complex Systems. Fractal and Multifractal Models for Interpretation of Experimental Data*. Cham, Switzerland: Springer-Verlag.
43. Teubner, M. and Strey, R. (1987). Origin of the scattering peak in microemulsions. *Journal of Chemical Physics* 87: 3195–3200.
44. Schubert, K.V., Strey, R., Kline, S.R., and Kaler, E.W. (1994). Small-angle neutron-scattering near Lifshitz lines - transition from weakly structured mixtures to microemulsions. *Journal of Chemical Physics* 101: 5343–5355.

45. Dehsorkhi, A., Castelletto, V., Hamley, I.W., and Harris, P.J.F. (2011). Multiple hydrogen bonds induce formation of nanoparticles with internal microemulsion structure by an amphiphilic copolymer. *Soft Matter* 7: 10116–10121.

46. Berk, N.F. (1987). Scattering properties of a model bicontinuous structure with a well-defined lengthscale. *Physical Review Letters* 58: 2718–2721.

47. Debye, P. and Bueche, A.M. (1949). Scattering by an inhomogeneous solid. *Journal of Applied Physics* 20: 518–525.

48. Debye, P., Anderson, H.R., and Brumberger, H. (1957). Scattering by an inhomogeneous solid. 2. The correlation function and its application. *Journal of Applied Physics* 28: 679–683.

49. Higgins, J.S. and Benoît, H.C. (1994). *Polymers and Neutron Scattering*. Oxford: Oxford University Press.

50. Klein, R. (2002). Interacting colloidal suspensions. In: *Neutrons, X-Rays and Light Scattering Methods Applied to Soft Condensed Matter* (eds. P. Lindner and T. Zemb). Amsterdam: Elsevier.

51. Hansen, J.P. and Macdonald, I.R. (2013). *Theory of Simple Liquids*, 4th edition. London: Academic Press.

52. Pedersen, J.S. and Gerstenberg, M.C. (1996). Scattering form factor of block copolymer micelles. *Macromolecules* 29: 1363–1365.

53. Pedersen, J.S. and Svaneborg, C. (2002). Scattering from block copolymer micelles. *Current Opinion in Colloid & Interface Science* 7: 158–166.

54. Pringle, O.A. and Schmidt, P.W. (1971). Small-angle x-ray scattering from helical macromolecules. *Journal of Applied Crystallography* 4: 290–293.

55. Kerker, M. (1969). *The Scattering of Light*. New York: Academic.

56. Porte, G. (2002). Surfactant micelles and bilayers: shapes and interactions. In: *Neutrons, X-Rays and Light Scattering Methods Applied to Soft Condensed Matter* (eds. P. Lindner and T. Zemb). Amsterdam: Elsevier.

57. Fournet, G. (1951). Etude theorique et experimentale de la diffusion des Rayons X par les ensembles denses de particules. *Bulletin de la Societe Francais de Mineralogie et Cristallographie* 74: 37–98.

58. Kratky, O. (1963). X-ray small angle scattering with substances of biological interest in diluted solutions. *Progress in Biophysics and Molecular Biology* 13: 105–173.

59. Knoll, W., Schmidt, G., Ibel, K., and Sackmann, E. (1985). Small-angle neutron-scattering study of lateral phase-separation in dimyristoylphosphatidylcholine cholesterol mixed membranes. *Biochemistry* 24: 5240–5246.

60. Glatter, O. (2018). *Scattering Methods and their Application in Colloid and Interface Science, Chap.5*. Amsterdam: Elsevier.

61. Rubinstein, M. and Colby, R.H. (2003). *Polymer Physics*. Oxford: Oxford University Press.

62. de Gennes, P.G. (1979). *Scaling Concepts in Polymer Physics*. Ithaca: Cornell University Press.

63. Schurtenberger, P. (2002). Static properties of polymers. In: *Neutrons, X-Rays and Light Scattering Methods Applied to Soft Condensed Matter* (eds. P. Lindner and T. Zemb), 259–298. Amsterdam: Elsevier.

64. Akcasu, A.Z., Benmouna, M., and Benoit, H. (1986). Application of random phase approximation to the dynamics of polymer blends and copolymers. *Polymer* 27: 1935–1942.

65. Alamo, R.G., Londono, J.D., Mandelkern, L. et al. (1994). Phase-behavior of blends of linear and branched polyethylenes in the molten and solid states by small-angle neutron-scattering. *Macromolecules* 27: 411–417.

66. Leibler, L. (1980). Theory of microphase separation in block copolymers. *Macromolecules* 13: 1602–1617.

67. Hamley, I.W. (1998). *The Physics of Block Copolymers*. Oxford: Oxford University Press.

68. Hamley, I.W. (2005). *Block Copolymers in Solution*. Chichester: Wiley.

# 2

# Data Analysis

## 2.1 INTRODUCTION

With sophisticated instrumentation, high-quality and information-rich small-angle scattering (SAS) data can be analysed to provide a wealth of information on ordering within bio- and nanomaterials. The first step, before modelling of the data, is the reduction and initial analysis of the data. This is the subject of Chapter 2.

This chapter covers data correction methods, treatment procedures including analysis of the Guinier and Porod regimes, and presentation of the data using Kratky and Zimm plots, as well as providing a list of software that is useful in data reduction and modelling. The post-reduction modelling of SAS data is discussed in the other chapters of this book, according to the technique employed.

This chapter is organized as follows. Section 2.2 considers the essential pre-measurement analysis of sample concentration and dispersity, particularly important for solution SAS on biomolecules. Section 2.3 provides an overview of steps in the SAS data processing pipeline. These are then discussed in detail in the following sections. Section 2.4 describes corrections for sample transmission, thickness, measurement time, and solid angle of scattering, while background corrections are discussed in Section 2.5. Section 2.6 is devoted to detector corrections, mask files, and integration. The analysis of anisotropic small-angle x-ray scattering (SAXS) data is discussed in Section 2.7, while the calibration of the $q$ scale is outlined in Section 2.8. For quantitative interpretation of data in terms of scattering densities (electron density for x-rays, scattering length density for neutrons) and

*Small-Angle Scattering: Theory, Instrumentation, Data and Applications*,
First Edition. Ian W. Hamley.
© 2021 John Wiley & Sons Ltd. Published 2021 by John Wiley & Sons Ltd.

determination of molar mass, the data should be placed in absolute units, as detailed in Section 2.9. Absorption effects are considered in Section 2.10, and smearing effects in Section 2.11, although this is mainly of concern to small-angle neutron scattering (SANS) measurements and is described in more detail in Section 5.5.1. Data checking procedures for synchrotron solution SAXS measurements are described in Section 2.12. The next sections are devoted to analysis of SAS data in different regimes. Section 2.13 discusses the high $q$ Porod regime, Section 2.14 concerns Kratky plots, and Section 2.15 is devoted to Zimm plots. The invariant and related information content from SAS measurements are discussed in Section 2.16. Section 2.17 includes a brief discussion of form factor fitting. The chapter concludes with Section 2.18 that presents a list of SAS software used for data analysis.

## 2.2   PRE-MEASUREMENT SAMPLE CONCENTRATION AND POLYDISPERSITY MEASUREMENTS

Typical solution SAXS and SANS measurements are performed on samples with concentrations in the range 1–20 mg ml$^{-1}$. As discussed in Section 2.12, measurement of several concentrations may be advantageous to check for aggregation and structure factor effects. For analysis of solution SAS data, sample concentration should be accurately determined before the measurement. For biomolecules, this can be done by UV absorbance measurements, which are particularly accurate for peptides/proteins containing an aromatic residue (since these give large defined absorbances measured at 280 nm); otherwise, the backbone absorbance band at 205 nm is measured [1]. Similarly, nucleotide absorbance measurements at 260 nm can be used to quantify concentration [2].

For polydisperse samples of biomacromolecules, the dispersity can be measured by size exclusion chromatography, gel electrophoresis, analytical ultracentrifugation, dynamic light scattering (DLS) or static light scattering (SLS), or mass spectrometry [3, 4].

## 2.3   OVERVIEW: DATA REDUCTION PIPELINE

There are a number of steps to process SAS data following both SAXS and SANS measurements, as shown schematically in Figure 2.1.

These corrections are discussed in turn in the following sections, with the exception of absolute intensity determination, which is discussed

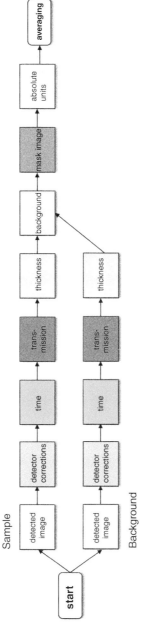

**Figure 2.1** Schematic pipeline for reduction of small-angle scattering data.

in Section 2.9. Software is available on SAXS and SANS beamlines or laboratory SAXS instruments to perform the data-processing steps shown in Figure 2.1. General programs to do this are also available as discussed in Section 2.18.

## 2.4 CORRECTIONS FOR SAMPLE TRANSMISSION AND OTHERS

The transmission of a sample is the ratio of the measured intensity after and before scattering by the sample, normalized by the same quantity measured without the sample [5–8]:

$$T = \frac{I_{a,s}}{I_{b,s}} \bigg/ \frac{I_{a,0}}{I_{b,0}} \tag{2.1}$$

Here a and b denote after and before the sample, and s and 0 label measurements with and without sample, respectively. As mentioned in Section 2.10, the transmission decreases exponentially with sample thickness:

On a SAXS beamline, the intensity (of a given data point) is corrected by the transmission factor $T$ and the incident flux, $F$:

$$I_{cor} = \frac{I}{TF} \tag{2.2}$$

The intensity is then further normalized for the sample measurement time $\tau$ and sample thickness $t$:

$$I'_{cor} = \frac{I_{cor}}{t\tau} \tag{2.3}$$

The errors on $I_{cor}$ are obtained from the root of the squared fractional errors in the terms in Eq. (2.2) and (2.3) [5]. In practice, all of these corrections can be done in one step by measuring the transmitted flux using a calibrated transmission monitor.

In SAXS experiments on synchrotron beamlines, the intensity before/after the sample can be monitored using inline semiconductor-based detectors, $p$-$i$-$n$ diodes, or ionization detectors. On SANS beamlines, the transmission is measured separately from the scattering curve, being obtained from the intensity over an integrated area over the central part of the detector (measuring direct beam) with an attenuator in place [7] (this method can also be used for SAXS, but the procedure is not commonly adopted). In $H_2O/D_2O$ mixtures, the transmission is a direct measure of the hydrogen content of a sample, since the attenuation is mainly due to the large, incoherent scattering from hydrogen atoms (Section 5.4). Thus, transmission measurements can be used to check the $H_2O/D_2O$ content in a sample compared to the corresponding buffer. For a series of $H_2O/D_2O$ contrasts, there is a linear relationship between $\log(T)$ and $D_2O$ content, so it is sufficient to measure transmission from two samples (e.g. pure $H_2O$ or pure $D_2O$) to determine the mole fraction of $D_2O$ in another solution of known transmission.

To compare SAS data measured for different sample-to-detector distances $(d_s)$ and acquisition times, it is necessary to normalize by the solid angle subtended by the detector pixel elements ($\Delta\Omega = p_1 p_2 d_s^2$, where the pixel size is $p_1 \times p_2$), and the incident number of photons (neutrons) $(i_0)$. The procedure is described elsewhere [9, 10].

## 2.5 BACKGROUND CORRECTIONS

Background refers to signal from many sources, including detector background due to naturally occurring and stray radiation (and dark current or detector noise in the case of older types of detectors such as multiwire detectors like charge-coupled devices (CCDs), Sections 3.4 and 3.5). As indicated in Figure 2.1, detector-specific background corrections should be separated (done before any normalization) from the background corrections with the beam. In the presence of the beam, there is background from stray scattering from optical elements in the beamline, windows, the sample holder, etc. Data can be corrected for the combined effect of these contributions to the sample by performing a measurement on an empty sample holder (capillary, quartz cell). This is not extremely reliable on synchrotron SAXS beamlines where small differences in the parameters of the sample holder (e.g. capillary wall thickness and diameter) from cell to cell lead to significant differences in the signal. It is less problematic on SANS beamlines where empty cell measurements are routinely used for background correction since the relative effect of variation in sample holder thickness on the lower neutron flux is proportionally smaller.

For SANS, the incoherent scattering background is very high in soft matter due to the high incoherent scattering from hydrogen as mentioned above. It is accounted for in the data-reduction process by examining the asymptotically flat intensity at the highest possible $q$ value. A similar procedure can be used for x-rays when an incoherent background is present due to fluorescence.

Finally, there is the background due to the solvent scattering from a sample in solution, which may be subtracted by measuring the SAS profile for the solvent separately and then subtracting this from the measured data. This works well on SANS beamlines and BioSAXS beamlines with fixed capillaries, but the same comment about empty cell variability holds otherwise for SAXS background correction [7].

Even after these correction procedures, the final corrected data will contain a background component that dominates at high $q$. This is due to inaccuracies in the data correction and/or the finite signal-to-noise level in the data and can be accounted for when fitting the data by using a minimal function such as a constant term to describe flat background scattering at high $q$ (correctly reduced ideal data should have data fluctuating around zero intensity at sufficiently high $q$). Problems of background subtraction are evidenced for example by apparent negative intensities (in oversubtracted data) or apparent increase in intensity with $q$ at high $q$ (undersubtraction). Wide-angle x-ray scattering (WAXS) data from solutions is even more sensitive to background subtraction than SAXS data and requires very careful sample preparation and data processing and often manual adjustment of the scale factor $\alpha$ in the expression for the corrected sample intensity:

$$I_{\text{sample}}(q) = I_{\text{measured}}(q) - (1 - \alpha)I_{\text{background}}(q) \qquad (2.4)$$

In the ideal case, $\alpha$ corresponds to the volume fraction of the constituents.

During a multiframe experiment, it is necessary to check the stability of the background signal (from buffer in the case of a BioSAXS experiment). This is typically done by measuring the background before and after a sample measurement, and checking it has not changed. This is to check for issues due to contamination, e.g. due to insufficient washing (residual sample) or radiation damage or problems with the buffer flow in the case of use of an SEC column in BioSAXS [11]. This is also discussed in Section 2.12. The development of variations in background or beam damage can be checked using CORMAP, which is a correlation map introduced for SAXS data checking [12], related to the variance–covariance of the measured intensities and their standard deviations. Specifically for $m$ frames of data containing $n$ data points $q_k$, the following reduced covariance parameter is computed:

$$r_{kl} = \frac{\sigma[I_{\text{exp}}(q_k), I_{\text{exp}}(q_l)]}{\sigma[I_{\text{exp}}(q_k)]\sigma[I_{\text{exp}}(q_l)]} \qquad (2.5)$$

where the diagonal variances on the measured experimental data points are given by

$$\sigma^2[I_{\text{exp}}(q_k)] = \frac{1}{m-1} \sum_{i=1}^{m} [I_{\text{exp}}(q_k)_i - \bar{I}_{\text{exp}}(q_k)]^2 \qquad (2.6)$$

Here, the bar denotes the mean intensity at a given $q_k$ data point, averaged over the $m$ frames. The off-diagonal (cross-correlation) covariances are given by

$$\sigma[I_{\text{exp}}(q_k), I_{\text{exp}}(q_l)] = \frac{1}{m-1} \sum_{i=1}^{m} [(I_{\text{exp}}(q_k)_i - \bar{I}_{\text{exp}}(q_k))(I_{\text{exp}}(q_l)_i - \bar{I}_{\text{exp}}(q_l))]$$
$$(2.7)$$

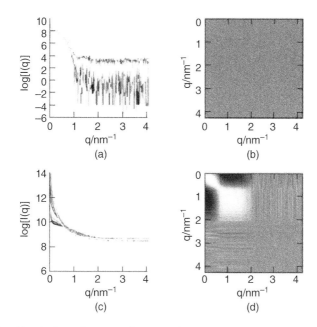

**Figure 2.2** SAXS data from 20-frame solution SAXS measurements showing CORMAP analysis for (a, b) good data from glucose isomerase, (c, d) data for a sample of lysozyme undergoing beam damage. SAXS data are shown in (a, c) and correlation maps in (b, d) with gray scale colours from −1 (black) to +1 (white). *Source*: From Franke et al. [12]. © 2015, Springer Nature.

The CORMAP reduced covariance (Eq. (2.5)) takes values in the range $-1 \leq r_{kl} \leq +1$. If there are no differences in data from frame to frame, then $I_{\exp}(q_k)$ and $I_{\exp}(q_l)$ are independent for all $l \leq k$, $l \leq n$, and $k \neq l$, and $r_{kl}$ takes random values. Figure 2.2 shows SAXS data and CORMAP plots for 20 frame files for the case of good data (no systematic variations) and data with beam damage showing clear features (positive and negative correlations) at low $q$.

## 2.6 DETECTOR CORRECTIONS, MASK FILES, AND INTEGRATION

For SAXS measurements, a variety of detector corrections must be performed, such as flat-field correction (which allows for variation in the sensitivity of individual pixels), dead-time correction (not necessary for modern hybrid pixel counting (HPC) detectors), and dark current and natural background correction, polarization corrections, and others, depending on the type of detector. Random spikes called 'zingers' due to cosmic rays or readout errors can occur, and are removed by 'dezingerization' as part of the detector correction procedure [4, 13].

On synchrotron beamlines, detector corrections are performed as part of the automated data reduction process and the user does not usually have to concern themselves with this. The dead or less-sensitive pixels (and nonpixel areas of detectors) are excluded from the subsequent data reduction process using 'mask files'. Detector corrections (and other data corrections) are discussed in detail elsewhere [5]. Information is available elsewhere for dead time corrections for SAXS [5] and SANS [7].

Stripes are observed in 2D detector images with photon-counting devices used on many SAXS beamlines such as Pilatus and other Dectris detectors, which are constructed from semiconductor modules and which have inactive areas due to gaps between the modules. Figure 2.3 shows an example of a 2D data file measured with such a detector, along with the necessary masking of the inactive strips necessary for data processing including reduction to a one-dimensional intensity profile. This reduction, or azimuthal integration, is done using software available at SAXS beamlines. The data are also commonly binned into courser $q$ bands in order to reduce noise associated with averaging over a limited number of pixels. Binning is usually done in equal steps of $\tan(2\theta)$ where $2\theta$ is the scattering angle [4, 13]. Binning of data and isotropic integration of data also introduce uncertainties into measured intensity profiles. The associated errors are discussed elsewhere [5, 13].

Masking is also performed for SANS data, but the 2D data does not have stripes. Typically, mask files for SAXS or SANS data analysis will be provided at the beamline, based on recent measurements with the same beamline configuration (i.e. same sample-detector distance, wavelength, etc.). The region of the beamstop (lower right in Figure 2.3) and dead or less-sensitive pixels are also excluded using mask files.

Detector-related corrections are generally done at SANS and SAXS beamlines automatically in the data reduction software. On SANS beamlines, the external background level of a detector can be measured by putting a neutron absorber such as cadmium (Table 2.1) at the sample position. Then the intrinsic background of the gas-filled multiwire detector, due to natural radiation and noise on the detector, can be measured. On modern SAXS beamlines, hybrid photon-counting detectors do not suffer from noise, but there will still be a background from natural radiation present at the beamline. This is usually measured by simply closing the beam shutter, but can also be done by masking the detector with an x-ray absorbent material.

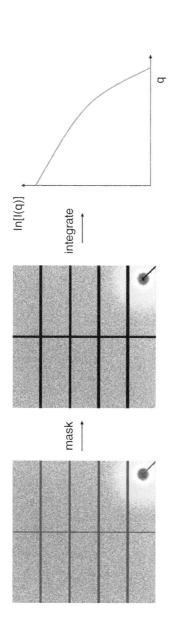

**Figure 2.3**  Masking of stripes in 2D SAXS data measured on a modern photon counting detector as part of the radial integration of the data. The beamstop is bottom right and the masked areas are blacked out.

The same corrections for measurement time, sample transmission, and thickness are made for the measured background sample prior to further reduction (Figure 2.1).

## 2.7  ANISOTROPIC DATA

For samples that undergo alignment, for example under flow, the raw 2D SAS pattern can be analysed for anisotropy and/or differences in sector- or box-integrated SAS data in different directions (typically orthogonal directions at $\phi = 0°$ and $90°$ azimuthal angles on the detector) as shown in Figure 2.4. Alignment may occur spontaneously under flow for some types of sample, or it can be induced in studies of soft materials using shear flow for example. Methods of shear alignment and so-called rheo-SAS are discussed in Section 3.12.2 and other alignment techniques (electric and magnetic fields and others) are also discussed in Section 3.12.

In solution SAXS on synchrotron beamlines or laboratory instruments, spontaneous orientation in 2D SAXS images is not very commonly observed, due to the dilute nature of solutions studied, and the globular nature of most biomacromolecules, which precludes flow alignment. However, spontaneous flow alignment in a capillary may be observed for sufficiently concentrated solutions of some polymers, liquid crystal phase-forming surfactants, copolymers, and amyloid fibre-forming peptides that exhibit a nematic phase with orientational ordering, as in Figure 2.4. The anisotropy can be analysed to provide information on the orientational order parameters of the molecules and/or anisotropic radii of gyration, which can be obtained from Guinier or Zimm plots of the data in orthogonal directions (an example of an anisotropic Zimm plot for a side-chain liquid crystal polymer is presented in Figure 2.16).

An aligned form factor pattern from rod-like structures will give an elliptical scattering pattern as in Figure 2.4. The degree of anisotropy can be related to orientation and shape anisometry. Different methods are available to extract orientational order parameters [15–19].

In systems showing interparticle interference, alignment gives rise to anisotropic structure factor features, as in liquid crystal phases including nematic and smectic phases. Analysis of orientation in SAS patterns is discussed in more detail elsewhere [18, 20–22].

## 2.8  CALIBRATION OF Q SCALE

On many synchrotron beamlines and reactor SANS beamlines, where the sample and detector position are fixed, the $q$-scale can be accurately

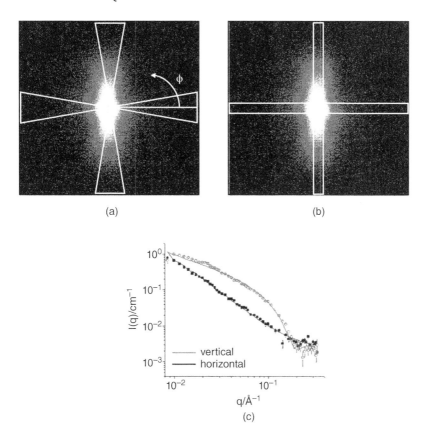

(a)                                    (b)

(c)

**Figure 2.4** Integrating anisotropic SAXS data: (a) Sector integration and definition of azimuthal angle $\phi$), (b) box integration, (c) sector integrated data along two orthogonal directions indicated (open symbols: measured data, lines: fits to anisotropic solid cylinder form factor). The data are for an amyloid forming peptide YYKLVFFC in aqueous solution that spontaneously underwent alignment in the capillary it was placed in, due to the formation of a nematic liquid crystal phase. *Source*: Adapted from Hamley et al. [14].

calculated geometrically knowing the sample-detector position, the wavelength, and the size of the pixels on the detector (for spallation SANS beamlines the $q$-scale is calculated from time-of-flight analysis as discussed in Section 3.6). For calibration purposes (always necessary on older beamlines and/or where the sample position is not fixed), collagen is often used to calibrate the SAXS data from systems with larger spacings (the $d$-spacing of collagen from wet rat-tail tendon, for example, is $d = 67\,\text{nm}$ [23], see Section 4.15) and it presents a series of well-defined evenly spaced Bragg peaks, which are oriented along the axis of the collagen. Silver behenate (silver salt of docosanoic acid, $C_{21}H_{43}COOH$), which has a lamellar structure with a $d = 58.380\,\text{Å}$, is used to calibrate many lab and

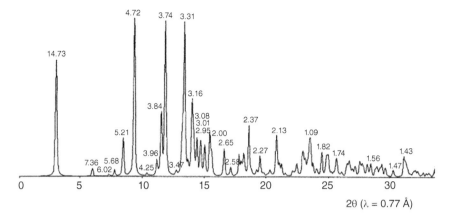

**Figure 2.5** Calculated WAXS pattern and peak indexation from para-bromobenzoic acid. *Source*: From Ref. [26]. © 2012, Elettra-Sincrotrone Trieste.

synchrotron systems with shorter camera lengths (cholesterol myristate is an alternative). WAXS detectors may be calibrated using NBS silicon powder (for peak indexation see Ref. [24]) or corundum (a National Institute of Standards and Technology Institute (NIST) standard exists for this crystalline form of aluminium oxide, $Al_2O_3$) or para-bromobenzoic acid, or with a sample of high-density polyethylene [25]. Figure 2.5 shows the WAXS pattern of para-bromobenzoic acid with peak indexation.

## 2.9  ABSOLUTE INTENSITY CALIBRATION

For x-rays, the intensity is usually put on an absolute scale with respect to the forward differential scattering cross-section for water $\left(\frac{d\sigma}{d\Omega}\right)_{q=0} = 1.65 \times 10^{-2}$ cm$^{-1}$, this being obtained from the relationship involving scattering length density and isothermal compressibility [7, 8, 27]:

$$\left(\frac{d\sigma}{d\Omega}\right)_{q=0} = \rho^2 k_B T \chi_T \tag{2.8}$$

Here, $\rho$ is the scattering length density (for x-rays) for water, $\rho = 9.452 \times 10^{10}$ cm$^{-2}$ at 20 °C [28], $(9.40 \times 10^{10}$ cm$^{-2}$ according to Ref. [29] and $9.469 \times 10^{10}$ cm$^{-2}$ in Table 6.1), and $\chi_T$ is the compressibility of water $(4.58 \times 10^{-10}$ Pa$^{-1}$ at 20 °C [27, 30], equivalent to $4.58 \times 10^{-4}$ cm$^3$ J$^{-1}$).

In either SAXS or SANS measurements, for a dilute solution of particles such as proteins, the effective molar mass, $M$, in g mol$^{-1}$ can be obtained from the forward scattered intensity, denoted $I(0)$ for convenience and placed on an absolute scale (in $cm^{-1}$) via the equation [8]

$$M = I(0)\frac{N_A}{c_p(\Delta\rho v_p)^2} \tag{2.9}$$

Here $N_A$ is Avogadro's number, $c_p$ is the particle concentration in $g\,cm^{-3}$, $v_p$ is the specific volume in $cm^3\,g^{-1}$, and $\Delta\rho$ is the average scattering density (scattering length density for neutrons or electron density for x-rays). In terms of number density of particles $n_p = c_pN_A/M$, Eq. (2.9) can be rearranged as

$$I(0) = n_p(\Delta\rho)^2 V^2 \tag{2.10}$$

The intensity can be put on an absolute scale either using primary standards of known calculable scattering power, or secondary standards which are pre-calibrated samples. The advantages and disadvantages of different approaches to absolute intensity scale normalization have been summarized, in the context of a lab SAXS instrument [27]. Analytical expressions based on the camera geometry and measured absorption coefficient are available to directly calibrate the intensity scales for Kratky SAXS cameras [18]. The uncertainties associated with absolute calibration of SAXS data are discussed elsewhere [5].

Secondary standards employed for absolute intensity calibration include Lupolen™ [5, 8, 18, 31], a type of low-density polyethylene or glassy carbon [5]. Lupolen samples with known absolute intensity may be passed on from one beamline to another. Lupolen has a broad peak around $q \sim 0.38\,nm^{-1}$ due to the lamellar repeat distance and the differential scattering cross-section for SANS of the peak has been well calibrated to be $\frac{d\sigma}{d\Omega} = 0.6\,mm^{-1}$ [10, 32].

For SANS, vanadium is used as a primary standard, since it produces strongly incoherent scattering, so $\frac{d\sigma}{d\Omega} = \Sigma_{inc}/4\pi$. Then the intensity measured in detector pixel $j$ can be written as [6]

$$I_j = \Phi_0\Omega_j\varepsilon(\lambda)Tt\Sigma_{inc}/4\pi \tag{2.11}$$

where the terms (also defined after Eq. (5.1)) are $\Phi_0$: incident flux, $\Omega_j$: solid angle subtended at pixel j, $\varepsilon(\lambda)$: detector efficiency at wavelength $\lambda$, and $t$: sample (in this case, vanadium) thickness. From Eq. (2.15), $T = I_j/I_{0,j} = \exp(-\mu t)$ and so [6]

$$I_j = \Phi_0\Omega_j\varepsilon(\lambda)\exp(-\mu t)t\Sigma_{inc}/4\pi \tag{2.12}$$

which can be used to correct the intensity of pixel j via the factor $\Omega_j\varepsilon(\lambda)$.

Using a polymer such as Lupolen as a secondary standard for SANS, the following equation is applied [7]:

$$\frac{\left(\dfrac{d\sigma}{d\Omega}\right)_{H_2O,eff}}{\left(\dfrac{d\sigma}{d\Omega}\right)_{polymer}} = \frac{I_{H_2O}}{I_{polymer}} = \frac{\overline{M}_w^{real}}{\overline{M}_w^{exp}} \tag{2.13}$$

where $\overline{M}_w^{real}$ is the actual polymer standard molar mass, and $\overline{M}_w^{exp}$ is that obtained experimentally in the limit $q \to 0$. For SANS, water is also used as a convenient secondary standard, since it presents a large flat (i.e. $q$-independent) scattering [7].

For SANS, it is also possible to directly put the data in absolute units by appropriate calibration of the detector using a strong scatterer. This enables the determination of the instrumental constant (cf. Eq. (5.1)):

$$C(\lambda) = \Delta\Omega \frac{I_{0,s}}{I_{0,i}} a D_s \varepsilon(\lambda) = I_{incident} \Delta\Omega \tag{2.14}$$

where $I_{incident}$ is the incident flux per second [7].

## 2.10   ABSORPTION

Absorption of x-rays leads to emission of electrons via the process of photo-electric absorption. Neutrons are absorbed by nuclei (leading in some cases to subsequent radioactive decay). In either case, the intensity decays exponentially as a function of depth, $z$:

$$I(z) = I_0 e^{-\mu z} \tag{2.15}$$

Here $I_0$ is the incident intensity and $\mu$ is the linear absorption coefficient, which is given by

$$\mu = \rho_a \sigma_a = \left(\frac{\rho_m N_A}{m}\right) \sigma_a \tag{2.16}$$

Here, $\rho_a$ is the number density of atoms, $\sigma_a$ is the absorption cross-section, $\rho_m$ is the mass density, and $m$ is the atomic mass. The absorption of a material may also be quantified as the mass absorption coefficient

$$\frac{\mu}{\rho_m} = \frac{N_A}{m} \sigma_a \tag{2.17}$$

Table 2.1 lists examples of x-ray and neutron absorption cross-sections and mass absorption coefficients for selected elements. These values are given at specific wavelengths, since the quantities are wavelength

**Table 2.1** Absorption cross-sections and mass absorption coefficients for x-rays (at $\lambda = 1.542$ Å) and neutrons ($\lambda = 1.8$ Å) [20].

| Element | Absorption cross-section $\sigma_{abs}$ $(10^{-24}$ cm$^2)$ | | Mass absorption coefficient $\mu/\rho_m$ (cm$^2$ g$^{-1}$) | |
|---|---|---|---|---|
|  | X-ray | Neutron | X-ray | Neutron |
| H($^1$H) | 0.655 | 0.332 6 | 0.391 | 0.199 |
| D($^2$H) |  | 0.000 51 |  | 0.000 15 |
| B | 41.5 | 767 | 2.31 | 42.7 |
| C | 89.9 | 0.003 5 | 4.51 | 0.000 18 |
| N | 173 | 1.90 | 7.44 | 0.081 7 |
| O | 304 | 0.000 19 | 11.5 | 0.000 007 2 |
| Na | 1140 | 0.530 | 29.7 | 0.013 9 |
| Al | 2220 | 0.231 | 49.6 | 0.005 16 |
| Cl | 6240 | 33.5 | 106 | 0.569 |
| Fe | 28 000 | 2.56 | 302 | 0.027 6 |
| Cd | 41 500 | 2520 | 222 | 13.5 |
| Gd | 105 000 | 48 890 | 403 | 187.2 |
| Pb | 79 800 | 0.171 | 235 | 0.000 50 |

*Source*: From Roe [20]. © 2000, Oxford University Press.

dependent, especially near x-ray absorption edges (in contrast the absorption cross-section for neutron is directly proportional to wavelength). The trend for these quantities to increase with atomic number is apparent for the case of x-rays whereas there is an irregular variation for neutrons. It is clear that certain elements are very high absorbers of neutrons (for example Cd or Gd), and are used as such as beamstoppers in SANS beamlines. Lead is commonly used as an absorber for x-rays.

The above analysis applies when the intensity attenuation is due to absorption, not scattering.

Empirically, the contribution of both absorption and scattering is included via an effective transmission $T_r = \exp(-\mu' t)$ where $\mu'$ is the effective absorption coefficient and $t$ is the sample thickness.

From Table 2.1, it may be noted that Cl has a significant absorption for x-rays; therefore, use of chlorinated solvents (such as guanidine hydrochloride in denaturing studies) should be avoided, where possible.

## 2.11  SMEARING EFFECTS

Smearing or resolution effects are usually considered in the modelling of the data, and are not performed as part of the data reduction process. Resolution effects arise from a variety of sources including the spread of neutron or x-ray wavelength, the finite resolution of the detector due to pixel size, the beam divergence, and beam width, resulting from collimation and sample thickness [5, 7].

Smearing effects are often neglected on synchrotron SAXS beamlines or lab pinhole SAXS cameras. Where these do need to be accounted for (in the case of high resolution measurements for example) the procedure steps in the calculation of instrumental resolution have been detailed [33]. It is important to consider instrumental resolution effects for SANS measurements [32, 34, 35], as discussed further in Section 5.5.1. Also, line-collimated SAXS instruments (Kratky cameras) require analysis of the smearing effect, which is intrinsic to this type of measurement, and this is preferably performed during the data analysis rather than attempting data desmearing [5].

## 2.12   SOLUTION SAXS DATA CHECKS

Synchrotron SAXS measurements on solutions typically operate in a high throughput mode, using fast multiframe measurements. It is good practice to monitor sets of multiframe measurements after they have been acquired in order to check data quality. The raw data are checked for beam damage (also discussed in Sections 2.5 and 3.8) by checking for changes in the intensity profile in successive frames, and if this is absent, the data are background subtracted. If beam damage is observed to develop after several frames, then these should be excluded and the averaging done with the ones where the scattering profile is unchanged (this assumes that a kinetic process is not being examined!). An alternative strategy in this case is to attempt to eliminate beam damage by reducing the exposure time, increasing the sample flow rate (or flowing the sample if it is not initially under flow) and/or attenuating the beam and/or reducing the path length through the sample as described for example in the context of BioSAXS capillary considerations in Section 3.8. However, there are limitations on reasonable sample flow rate, since if this is excessive, bubble formation will occur, which leads to problems with forward scattering, described below. Figure 2.6 shows a comparison (Figure 2.6a,b) of SAXS data measured over $5 \times 50$ ms frames for a solution of lysozyme without beam damage for a sample under flow through a capillary and a sample with beam damage, which occurred when flow was stopped [36]. This beam damage can be eliminated by attenuating the beam (with aluminium foils) as shown in Figure 2.6c,d.

Another strategy to reduce beam damage in susceptible samples is to add free radical scavengers such as dithiothreitol (DTT), TCEP

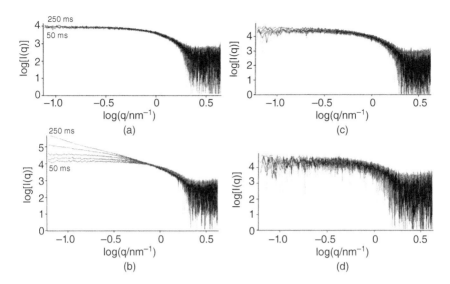

**Figure 2.6** SAXS data measured for lysozyme (4 mg ml$^{-1}$ in buffer, $5 \times 50$ ms frames) with full beam flux $5.1 \times 10^{12}$ photons s$^1$. (a) With sample flow enabled (30 μl s$^{-1}$), (b) in the absence of flow. Beam damage is manifested in the latter case by an increase in forward scattering. Amelioration of beam damage by attenuation using Al foils to reduce the flux to (c) $7.3 \times 10^{11}$ photons s$^{-1}$, (d) $1.8 \times 10^{11}$ photons s$^{-1}$. *Source*: From Jeffries et al. [36]. © 2015, International Union of Crystallography.

[tris(2-carboxyethyl)phosphine] or ascorbic acid, which scavenge reactive oxygen species, or polyols, which react with hydroxyl radicals and which reduce protein-protein interactions, relevant to suppression of aggregation in protein BioSAXS [37–39]. Figure 2.7 shows examples of the reduction of beam damage achieved using such radical scavengers in SAXS data from the enzyme RNAse in buffer solution [36]. In this study, the authors also calculated the critical x-ray dose (in kilograys, kGy, equivalent to energy absorbed in kJ kg$^{-1}$) that causes a defined change in radius of gyration values [36]. Radiation dose-dependent aggregation was examined previously [37]. Severe beam damage of the sample (when not under flow) was observed after 30 ms, which was ameliorated using radical scavengers. Tris [tris(hydroxymethyl)amino methane] or HEPES [4-(2-hydroxyethyl)-1-piperazineethanesulfonic acid] in buffers can also act as radical scavengers.

Cryogenic cooling, which is used in protein crystallography to reduce beam damage, is generally not suitable for BioSAXS, since the interest is in measurements on the samples in aqueous solutions under native conditions. However, this method, so-called cryo-SAXS, has been demonstrated [40].

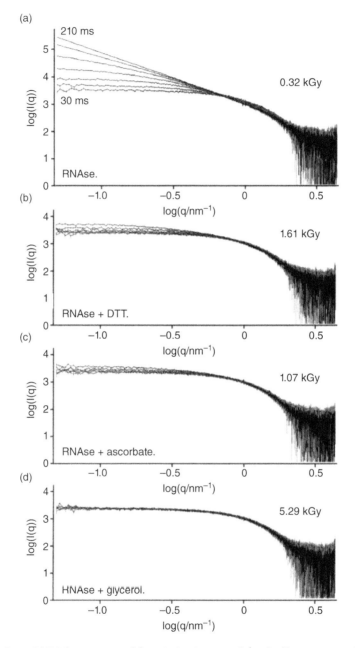

**Figure 2.7** SAXS data measured for RNAse (10 mg ml$^{-1}$ in buffer, $7 \times 30$ ms frames) with full beam flux $5.1 \times 10^{12}$ photons s$^{-1}$ in the absence of flow. (a) Without radical scavengers, beam damage is observed after 30 ms. Amelioration of beam damage by addition of (b) 1 mM DTT, (c) 1 mM ascorbate, (d) 5% v/v glycerol, in this case no beam damage is observed up to 300 ms. The critical x-ray dose (in kilograys, kGy, equivalent to energy absorbed in kJ kg$^{-1}$) that causes a defined change in radius of gyration values are indicated. *Source*: From Jeffries et al. [36]. © 2015, International Union of Crystallography.

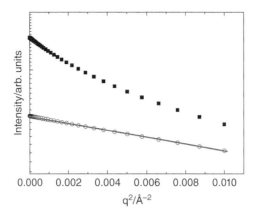

**Figure 2.8** Guinier plots (data on a logarithmic intensity scale) showing a Guinier regime for the lower data set (straight red line) and upward curvature for a system with aggregation (upper points).

Aggregation into multimers (dimers and higher structures), clusters or precipitation is problematic for samples that are not sufficiently dilute, these are manifested in increased scattering at very low $q$. This is shown, for example, in Guinier plots (Section 1.4) as presented in Figure 2.8. The lower plot shows a straight-line behaviour as expected, from which $R_g$ and $I(0)$ can be obtained in the typical Guinier regime, which is valid in the range $qR_g < 1$ [8, 20, 41]. However, the upper curve shows an upturn in the scattering at low $q$ (i.e. positive curvature), which is indicative of the extent of aggregation. A small upturn at low $q$ may be ameliorated by sample dilution (the data in Figure 2.8 shows a more major upturn). A series of Guinier plots at different concentrations can reveal the presence of any concentration-dependent aggregation processes.

During capillary flow delivery in BioSAXS, bubbles in the sample (even microbubbles) or the meniscus of the sample passing the x-ray beam can cause large changes in scattering profiles. During delivery of sample, the presence of bubbles or meniscus causes 'random' changes in the intensity over frames. This is illustrated in Figure 2.9. The first three frames of raw data show very much higher scattering than the remaining seven frames (which give the expected signal for this sample), which is due to bubbles and/or meniscus in the beam. Therefore, in the subsequent averaging the first three frames were excluded.

Bubbles, if present in the sample as prepared, can be removed by freeze-thawing or by vacuum treating to remove microbubbles in more viscous samples. However, freeze/thaw cycles can lead to aggregation

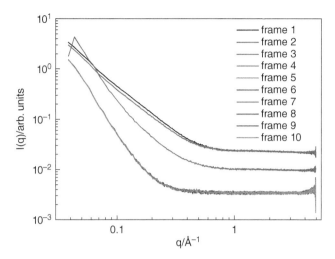

**Figure 2.9**   Example of raw synchrotron BioSAXS data showing artefacts in the first 3 frames of 10. *Source*: Unpublished data from ESRF.

of biomacromolecules. Bubbles may also be produced by beam damage (liberation of oxygen), methods to reduce which are discussed in Section 3.8.

A concentration series may be measured to evaluate the effects of inter-particle interference and the data for sufficiently low concentration, which do not show concentration-dependent low $q$ scattering may be selected for further analysis. Aggregation effects can be circumvented by thorough filtering of the sample or separation through SEC-SAXS. Centrifugation or ultracentrifugation may alternatively be performed. As mentioned above, aggregation effects are probed from Guinier plots. These are expected to be more important as concentration increases. Samples of self-assembling molecules such as surfactants, amphiphilic copolymers, lipidated peptides, etc. have an intrinsic aggregation propensity above a critical concentration, and it may be of interest to study this, rather than attempt to eliminate it by filtration and related methods, which in any case are difficult to apply to separate nanoscale assemblies.

Use of very dilute samples leads to poor signal-to-noise so it can be a matter of balancing sufficient dilution to avoid aggregation with sufficiently high concentration to obtain decent signal. Typically on synchrotron BioSAXS beamlines, reliable data requires a sample concentration in excess of $1 \, \mathrm{mg\,ml^{-1}}$ or more. If the concentration is too high, then inter-particle interference (structure factor) effects become apparent. These are manifested by the presence of a low $q$ plateau or peak, as illustrated in Figure 2.10. Structure factors are discussed in more detail in Section 1.6.

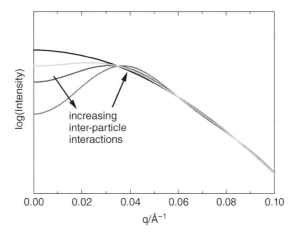

**Figure 2.10** Schematic showing effect of interparticle interference (structure factor) at low $q$. The black line corresponds to the pure form factor, measurable in a sufficiently dilute solution, the coloured lines show the effect of increasing structure factor.

It is common to measure data at a series of concentrations $c$ and if $I(q)/c$ curves superpose, this can be taken as an indication of a lack of interparticle interference effects. Figure 2.11 shows a counterexample of non-superposing $I(q)/c$ curves for a sample that undergoes concentration-dependent aggregation, in this case the surfactant-like peptide self-assembles into fibrils as

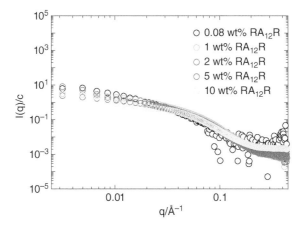

**Figure 2.11** Non-superposition of $I(q)/c$ curves for solutions of a surfactant-like peptide (arginine-alanine$_{12}$-arginine). *Source*: Adapted from Charlotte Edwards-Gayle, PhD Thesis, University of Reading, 2020.

concentration increases. In some studies, it is of interest to measure the structure factor, when information on inter-particle interactions is sought.

Since structure factor or aggregation effects influence the intensity at low $q$, these can introduce artefacts into the determined pair distance distribution function $p(r)$ (Eq. (1.19)) as illustrated in Figure 2.12 [42]. An initial examination of SAS data often involves a Guinier plot (as discussed above). Good data (Figure 2.12a) leads to a linear Guinier plot (Figure 2.12d) and a well-defined $p(r)$ that does not change when the selected $D_{max}$ is changed (Figure 2.12g,j). This is in contrast to the behaviours exhibited for samples with inter-particle interference, which show nonlinear Guinier

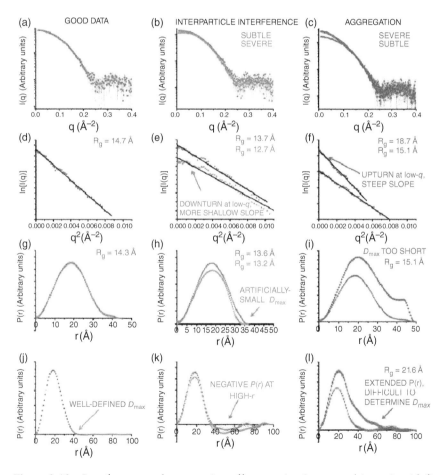

**Figure 2.12** Interference and aggregation effects on (a–c) scattered intensity, (d–f) Guinier plots, (g–i) pair distance distribution functions $p(r)$ using a low $D_{max}$ cut-off, (j–k) $p(r)$ over an extended range of $r$ (higher $D_{max}$). *Source*: From Jacques and Trewhella [42]. © 2010, John Wiley & Sons.

plots (Figure 2.12e) or for samples with aggregation, which causes changes in slope (Figure 2.12f). Such samples also produce unreliable pair distance distribution functions, the shape of which changes with the chosen $D_{max}$ (compare parts h with k, and i with l in Figure 2.12) and/or that show artefacts in their shape (Figure 2.12i).

Mismatched buffer can also be problematic in SAXS data analysis, i.e. variation between the sample and the background buffer and/or between different samples dissolved in the buffer [3]. This can be avoided by very careful buffer preparation, with dialysis. Also, in our experience, the presence of organic solvent residues (e.g. trifluoroacetic acid (TFA) or methanol from peptide synthesis or dissolution or other solvents from sample synthesis) can lead to problems with buffer subtraction. Some organic solvents such as methanol that are immiscible with aqueous buffers undergo phase separation and have large (and not very reproducible) backgrounds.

Useful summaries of data-quality checks for BioSAXS measurements, for example, on proteins in solution, are available [4, 43, 44]. As well as the analysis of data from solutions of proteins, these procedures may also be used for other SAXS studies on solutions of soft materials, for example.

After these checks, BioSAXS data may be used to obtain $R_g$, the radius of gyration, from a Guinier plot (Section 1.4) and the molecular weight (molar mass) from $I(0)$ (Section 2.9). The pair distance distribution function $p(r)$ may then be obtained as discussed in Section 4.6.1, and the flexibility of the biomacromolecule may be assessed by Porod-Debye and/or Kratky plots as described in the following sections. Zimm plots may be used instead of Guinier plots, and these are described in Section 2.15.

## 2.13   POROD REGIME

The Porod or Porod-Debye regime (derived by Porod in 1951 [45] and Debye et al. in 1957 [46]) describes the scattering at high $q$, providing information on the shape and surface-to-volume ratio of particulate/globular scattering objects [46].

For systems composed of particles with sharp interfaces, the scattering at high $q$ is given by [8, 20, 41, 46, 47]:

$$\lim_{q \to \infty} I(q) = \frac{2\pi(\Delta\rho)^2}{q^4} \frac{S}{V} \qquad (2.18)$$

Here $S/V$ is the specific surface (surface area/volume). This expression can be obtained from Eq. (1.15) and (2.28), taking a series expansion of $\sin(qr)/qr$ and retaining the leading term [47]. The so-called Porod law is the scaling $I \sim q^{-4}$ observed for compact particles in the two-phase approximation of a sharp interface between the particle and the solvent [8, 20, 41, 43, 46, 47].

For a two-phase system, the average length of a chord (a chord is a line in an arbitrary direction, joining points on the surface, the average is the average thickness of that phase) of one phase is

$$\langle l_x \rangle = 4 \frac{V}{S} \phi_x \qquad (2.19)$$

where $\phi_x$ labels the volume fraction of phase 1 or 2. This can be computed if $(\Delta \rho)$ is known, using Eq. (2.18). Porod's length of inhomogeneity, $l_P$, is the geometric mean given by [20, 47]

$$\frac{1}{l_P} = \frac{1}{\langle l_1 \rangle} + \frac{1}{\langle l_2 \rangle} \qquad (2.20)$$

The quantity $l_P$ is a measure of the average size of heterogeneities in the system.

Plots of $I(q)q^4$ vs. $q$ can be used to highlight features of scattering at high $q$ that deviate from the Porod law prediction, due to the presence of smooth interfaces or particle curvature (or fractal structure). Porod-Debye plots are commonly employed to obtain the specific surface from a plot of $Iq^4$ versus $q^4$. A Porod-Debye plot will show a Porod plateau, which from Eq. (2.18) can be used to determine $S/V$. Figure 2.13 shows examples of Porod-Debye plots from a protein showing a plateau in the folded state and in an unfolded conformation where no plateau is present [48]. Thus, the Porod-Debye plot also provides information on the folding state of macromolecules (cf. Kratky plots, Section 2.14). It has been shown that these plots provide insight into the flexibility of biological macromolecules in solution, and are more informative than Kratky plots, which are subject to inaccuracies and uncertainties associated with background subtraction and the limited range of $q$ over which the data is measured [48, 49]. The Porod-Debye analysis applies to a lower $q$ range of data, and is less subject to problems with background subtraction discussed in Section 2.12.

## 2.14   KRATKY PLOTS

Kratky plots are graphs of $q^2 I(q)$ versus $q$, which emphasize any deviation from $q^{-2}$ scattering of random coils. Figure 2.14 shows Kratky plots for a polymer coil, which is characterized by a plateau compared to that for a star polymer (with eight arms), which has a maximum. Kratky plots also highlight the difference between folded proteins with a compact

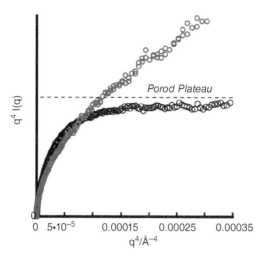

**Figure 2.13** Example of Porod-Debye plot for a hybrid protein (RAD51AP1-maltose binding protein conjugate) in folded state (black circles and Porod plateau observed) and unfolded state (red circles). *Source*: From Rambo and Tainer [48]. © 2011, John Wiley & Sons.

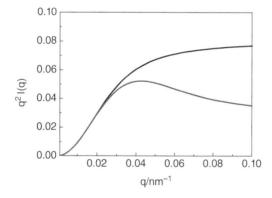

**Figure 2.14** Examples of Kratky plots: black line: polymer coil (calculated from the Debye function shown in Figure 1.21); red line: star polymer with eight arms with $R_g = 50\,\text{Å}$.

conformation, which show a maximum in the Kratky plot and intrinsically disordered proteins, which have a coil-like conformation, the SAS profile of which exhibits no maximum in a Kratky plot.

Kratky plots for random coil or unstructured chains (polymers or peptides) can be fitted with worm-like chain models using a variety of expressions corresponding to different limits (of chain length, persistence length, etc.) [51–57]. These have quite complicated analytical expressions (some SAXS form factor fitting software such as SASfit, Table 2.2, is able to fit worm-like chain models). An approximate formula is available for the infinite Kratky-Porod chain of contour length $L$ in the limit $qb > 4$, where $b$ is the statistical segment length that is twice the Kuhn persistence length [43, 54, 58]. The equation is [43, 54, 55, 59]:

$$P(q) = \frac{\pi}{qL} + \frac{2}{3q^2bL} \tag{2.21}$$

This expression is valid for wormlike chains without excluded volume. Alternative approximations are available in this case, as are approximations for wormlike chains without excluded volume, that are valid over wider $q$ ranges [60]. Other earlier numerical interpolation formulae for wormlike chain models have also been presented [61, 62]. An analytical form factor is also available for a star molecules in which each branch is a wormlike chain [63–65].

On the other hand, for finite Kratky-Porod chains at low $q$ ($qb < 2$), the form factor can be approximated as [43, 54, 59]

$$P(q) = \left(\frac{2}{x^3}\right)(e^{-x} - 1 + x) + \frac{2}{15u}\left[4 + \frac{7}{x} - (11 + \frac{7}{x})e^{-x}\right] \tag{2.22}$$

Here, $x = \frac{q^2 L b}{3}$ and $u = \frac{L}{b}$.

Figure 2.15 shows examples of Kratky plots measured by SANS for homopolymers with different tacticity. The observed maxima highlight the deviations from Gaussian coil behaviour due to excluded volume effects.

A Kratky-Porod plot, used for two-dimensional disc and lamellar systems, involves plotting $q^2 I(q)$ versus $q^2$ instead of $q$, according to the equation [47, 68, 69]

$$I(q) = I(0)\exp(-R_c^2 q^2) \tag{2.23}$$

where $R_c$ is the radius of gyration of the cross-section of disc/layer. The thickness $T$ of the disc/layer is then $T = \sqrt{12}R_c$ (Section 1.4).

Limitations in the use of Kratky plots, particularly to analyse SAXS data from macromolecules in solution, are discussed in Section 2.13.

## 2.15   ZIMM PLOTS

For soft materials such as polymers, Zimm plots are sometimes used instead of Guinier plots to provide the radius of gyration. They are commonly employed in the analysis of light scattering data, but are also used for SANS or SAXS data analysis. Zimm plots can also be used to determine the molar

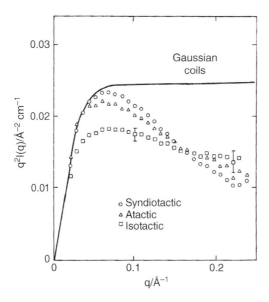

**Figure 2.15** Kratky plots of SANS data for poly(methyl methacrylate) (PMMA) with the tacticities indicated and the expected behaviour for Gaussian coils shown schematically as the solid line. *Sources*: Based on Bates [66] and O'Reilly et al. [67].

mass from intercepts (Eq. (2.24)), provided the intensity is on an absolute scale. The equation for a Zimm plot for a dilute sample takes the form [20, 22, 70]:

$$\frac{Kc}{I(q)} = \frac{1}{\overline{M}_w}\left(1 + \frac{q^2 R_g^2}{3} + \ldots\right) \tag{2.24}$$

Here, $K$ is an instrument-dependent constant, $c$ is the sample concentration, $\overline{M}_w$ is the weight-average molar mass, and $R_g$ is the radius of gyration. For a concentration series the Zimm plot equation is modified to

$$\frac{Kc}{I(q)} = \frac{1}{\overline{M}_w}\left(1 + \frac{q^2 R_g^2}{3} + \ldots\right) + 2A_2 c \tag{2.25}$$

Here $A_2$ is the second virial coefficient. A series of lines are plotted at different concentrations, displaced horizontally by fixed terms $k'c$, where $k'$ is a constant. This equation is used to obtain $A_2$, for example, for interacting proteins, in the case that the structure factor is significant. In both Eq. (2.24) and (2.25), the higher-order terms are usually neglected. The Zimm equation, Eq. (2.25) (in the $q = 0$ limit) can be derived from a virial expansion of the osmotic pressure as a function of concentration, using Eq. (1.114), which relates $I(0)$ to the concentration gradient of the osmotic pressure [20, 70, 71].

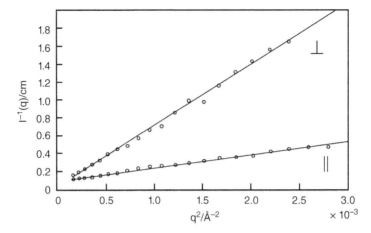

**Figure 2.16** Example of Zimm plots obtained from a SANS study of the backbone conformation of a liquid crystal polymer, using a mixture of polymers with deuterated and protonated backbone chains. The polymer was aligned in a magnetic field leading to anisotropic SANS with different slopes for the radii of gyration parallel and perpendicular to the magnetic field. *Source*: Redrawn from Hamley et al. [72].

The series expansions of the Guinier law equation and that for a Zimm plot provide $R_g$ from the first-order terms in the respective Eqs. (1.24) and (2.24); however, the higher-order terms are different so $R_g$ values obtained from the two types of plot may differ.

An example of the use of a Zimm plot to determine the radius of gyration for a blend of deuterated and protonated versions of the same polymer is illustrated in Figure 2.16. As discussed in Sections 5.7 and 5.8, this is a method to isolate single molecule chain scattering. Two radii of gyration were obtained from anisotropic SANS data from an aligned sample of a liquid crystal polymer blend [72]. The slopes of the lines in Figure 2.16 provide $R_g$ according to Eq. (2.24) and there must be a common intercept since the polymer has a unique molar mass (the intercept in Eq. (2.24) is equal to $1/\overline{M}_w$ where $\overline{M}_w$ is an effective weight-average molar mass).

## 2.16 INVARIANT AND RELATED INFORMATION CONTENT FROM SAS MEASUREMENTS

The integrated total intensity scattered over $q$ (over the reciprocal space on the surface of a sphere) is a conserved quantity, which only depend on the scattering density of the system (with respect to the background).

The so-called invariant can be written for a system of particles of volume $V_p$ as [8, 20, 47, 73]:

$$Q = \int_0^\infty I(q)q^2 dq = 2\pi^2 (\Delta\rho)^2 V_p \tag{2.26}$$

For a two-phase system (common in soft materials, for example, polymer blends or semicrystalline polymers) with volume fraction of one phase $\Phi$ (and scattering density difference between phases ($\Delta\rho$), the invariant is [20, 47, 73]

$$Q = \int_0^\infty I(q)q^2 dq = 2\pi^2 \Phi(1 - \Phi)(\Delta\rho)^2 \tag{2.27}$$

From Eq. (2.27), along with the expression for the forward scattering written in the form $I(0) = (\Delta\rho)^2 V_p^2$, the particle volume is given by

$$V_p = 2\pi^2 I(0)/Q \tag{2.28}$$

The particle volume can also be obtained from the integration of the auto-correlation function or pair distance distribution function:

$$V_P = \int_0^{D_{max}} 4\pi r^2 \gamma(r) dr \tag{2.29}$$

or

$$V_p^2 = \int_0^{D_{max}} p(r) dr \tag{2.30}$$

Another limiting behaviour is that of the correlation function, which can be expanded at small $r$ as

$$\gamma(r) = 1 - \frac{r}{4}\frac{S}{V} \tag{2.31}$$

The specific volume can thus be obtained from the slope of a plot of $\gamma(r)$ vs. $r$ at small $r$.

## 2.17 FORM FACTOR FITTING

For protein (and other biomacromolecule) solution SAS, it is usual to model the data by methods described in Sections 4.6 and 5.6.5 using conformations where available from x-ray diffraction, NMR, simulations, or based on *ab initio* methods. However, for self-assembled systems and nanostructures, it is usual to model SAXS data from sufficiently dilute solutions (i.e. those for which structure factor effects can be neglected) by fitting form factors. Section 1.7 provides examples of the equations, limiting behaviours, and the effects of polydispersity. The form factor model selected will usually be guided by prior information on the likely nanostructure in the system, based on other data such as microscopic characterization of the morphology or data from similar or related systems. Also, in some cases, constraints can be imposed on the model form factor using known molecular dimensions. For example, the radius of a micelle or cylindrical fibril cannot exceed the length of an amphiphilic molecule and a lamellar or vesicle bilayer thickness cannot exceed twice the extended molecular length. Measurements of intensity on an absolute scale also provide constraints on form factor modelling, since only electron or scattering length density values so obtained that are physically realistic are acceptable. The form factor model parameters can be used to construct the real space scattering density profile.

Several programs to perform form factor fitting are available (listed in Table 2.2). These use different least-squares fitting procedures to perform the fits. More details on least-squares fitting methods can be found elsewhere [74, 75]. The quality of fit is usually assessed by the chi-squared ($\chi^2$) measure:

$$\chi^2 = \sum_{i=1}^{N} \left( \frac{I_{\text{exp}}(q_i) - I_{\text{model}}(q_i)}{\sigma_i} \right)^2 \tag{2.32}$$

This function represents the difference squared between the experimental and model calculated intensity for $N$ data points with measured uncertainties $\sigma_i$. The input for most form factor fitting programs is a table of data with columns $q_i$, $I_{\text{exp}}(q_i)$, and $\sigma_i$, which are produced after data reduction at the beamline (or laboratory instrument).

The reduced chi-square value, defined as [74, 75]

$$\chi_r^2 = \frac{\chi^2}{N - M} \tag{2.33}$$

is also useful. Here $N - M$ is the number of degrees of freedom ($M$ is the number of model parameters). A perfect fit has $\chi_r^2 = 1$. It is also a good idea to check the residuals (difference between measured and model intensity profile or the ratio of these quantities) to examine any possible systematic differences in particular $q$ ranges between the data and the model. It should be noted that in many instances, a unique form factor fit cannot be obtained (not least due to the information content of the measurement discussed for example in Section 4.6.2). A high-quality fit that is consistent with molecular parameters and/or other measurements is what is aimed for.

A good fitting routine will also provide uncertainties on the fitted parameters. It should be noted that the quality of fit should be shown by plotting the intensity on a logarithmic scale (and ideally with $q$ also on a logarithmic scale, in a double logarithimic plot). A linear plot of fitted intensity versus $q$ should be treated with scepticism, since this form of plot can hide low-quality

**Table 2.2** SAS Software and online calculators.

| Name | Summary of capabilities | Web address | Reference |
| --- | --- | --- | --- |
| AquaSAXS | Calculates solution scattering from a pdb or similar file | https://omictools.com/aquasaxs-tool (part of OMIC tools, not free access) | [76] |
| ATSAS | Wide-ranging collection of routines to analyse and model SAXS data including $p(r)$ determination and many routines to model SAXS esp. from protein structures (discussed further in Section 2.12) | https://www.embl-hamburg.de/biosaxs/software.html | [77] |
| AXES | Calculation of solution SAXS profiles from pdb file | https://spin.niddk.nih.gov/bax/nmrserver/saxs1 | [78] |
| BayesApp | Website for indirect Fourier transformation of SAS data | https://somo.chem.utk.edu/bayesapp | [79] |
| BILBOMD | Conformational sampling for flexible proteins | https://bl1231.als.lbl.gov/bilbomd | [80] |
| BioXTAS RAW | SAXS data reduction and analysis including indirect Fourier transform, $I(0)$, Guinier fit and some ATSAS functions | https://bioxtas-raw.readthedocs.io/en/latest/index.html | [13] |
| BornAgain | Python-based software to model GISAXS data within DWBA | http://www.bornagainproject.org | [81] |
| CCP13 | Legacy software for fibre diffraction/SAXS data reduction and analysis including one-dimensional correlation function calculation (CORFUNC) | www.diamond.ac.uk/Instruments/Soft-Condensed-Matter/small-angle/SAXS-Software/CCP13.html | |
| D+ | Models BioSAXS from subunit assemblies (viruses, microtubules, etc.) | https://scholars.huji.ac.il/uriraviv/software/d-software | |
| DAWN | Comprehensive suite of data reduction tools. Not user friendly | https://dawnsci.org/downloads | |

(continued)

Table 2.2   (*continued*)

| Name | Summary of capabilities | Web address | Reference |
|---|---|---|---|
| DPDAK | For data reduction of large datasets of 2D SAXS and GISAXS data | https://confluence.desy.de/display/DPDAK | [82] |
| FIT2D | Legacy software with data reduction tools and peak fitting | http://www.esrf.eu/computing/scientific/FIT2D | [83] |
| FitGISAXS | Fitting of GISAXS data, requires commercial IgorPro software | http://gisaxs.com/index.php/Software | [84] |
| FoXS/MultiFoXS/FoXSDock | Calculation of solution SAXS profiles from pdb files for BioSAXS protein structure studies | http://modbase.compbio.ucsf.edu/foxs/about.html https://modbase.compbio.ucsf.edu/multifoxs | [85, 86] |
| GENFIT | Modelling of one-dimensional SAS data | https://sites.google.com/site/genfitweb | [87] |
| GIFT | Generalized indirect Fourier transformation for $p(r)$ determination | Believed to be available on request from Prof. O. Glatter (TU Graz, Austria) and supplied by Anton-Paar with their SAXS instruments | [88] |
| HipGISAXS | High-performance (massively parallel) software for simulating GISAXS data | https://hipgisaxs.github.io | [89] |
| IRENA | Package available for the commercial IgorPro software for SAS data reduction, analysis, and fitting | https://saxs-igorcodedocs.readthedocs.io/en/stable/Irena/Introduction.html | [90] |
| IsGISAXS | Analysis of grazing incidence small-angle x-ray scattering from nanostructures | http://www.insp.jussieu.fr/oxydes/IsGISAXS/isgisaxs.htm | [91] |

| Name | Description | URL | |
|---|---|---|---|
| MANTID | SANS data reduction and analysis. Used for other neutron and muon instruments also | https://www.mantidproject.org/Main_Page | [92] |
| McSAS | Calculates free form size distributions from SAS data using a Monte Carlo modelling method | https://bitbucket.org/pkwasniew/mcsas/src/master | [93] |
| MULCH | Calculates neutron and x-ray scattering length densities and contrast and $R_g$ for mixtures | http://smb-research.smb.usyd.edu.au/NCVWeb/input.jsp | [94] |
| NANOCELL | NANOCELL Simulates 2D diffraction patterns from single-crystals for GISAS, using the DWBA | https://sites.google.com/uw.edu/hillhouse/software | [95] |
| NCNR | Data reduction and fitting of SANS and USANS data, using commercial IgorPro software | https://www.nist.gov/ncnr/data-reduction-analysis/sans-software | [96] |
| NIKA | 2D to 1D data reduction tools for USAXS, SAXS, and WAXS data. Requires IgorPro commercial software | https://usaxs.xray.anl.gov/software/nika | [97] |
| NIST scattering length density, absorption, and neutron activation calculator | Calculations for elements for SAXS and SANS | https://www.ncnr.nist.gov/resources/activation | |
| NIST neutron reflectivity calculator | Calculates specular reflectivity for slab density model | https://www.ncnr.nist.gov/instruments/magik/calculators/reflectivity-calculator.html | [98] |
| PySAXS | Open Source Python package and GUI for SAXS data treatment | http://iramis.cea.fr/en/Phocea/Vie_des_labos/Ast/ast_sstechnique.php?id_ast=1799 | |

*(continued)*

Table 2.2  (continued)

| Name | Summary of capabilities | Web address | Reference |
|---|---|---|---|
| SASET | For data series analysis of isotropic and anisotropic data | 'Available on request from authors' | [19] |
| SASfit | Very useful software to fit a very wide range of form and structure factors for soft materials with a good fitting algorithm | https://www.psi.ch/en/sinq/sansi/sasfit | [99] |
| SAS Portal | Wide compilation of software for SAXS data reduction, analysis, and fitting | http://smallangle.org/content/Software | |
| SASSIE | Suite of programs to calculate SAS intensity profiles from atomistic configurations of proteins from pdb files or MC/MD simulation files from tools within the package. Also includes a scattering density and $I(0)$ calculator. | https://sassie-web.chem.utk.edu/sassie2/docs/sassie_docs.html | [100] |
| SASTBX | Small angle scattering toolbox including data reduction, PDDF calculation, and low resolution modelling | http://liulab.csrc.ac.cn/sastbx/pregxs.html Documentation: https://sastbx-document.readthedocs.io/en/latest# | [101] |
| SASview | Data analysis and visualization and with form factor fitting | https://www.sasview.org | |
| SAXSMoW | Web based calculation of $R_g$, molecular weight and Guinier, Kratky and Porod plots, etc. from SAXS intensity profiles | http://saxs.ifsc.usp.br | [102] |
| SAXSUtilities | Data reduction and visualization including 2D data | http://www.sztucki.de/SAXSutilities | |

| Name | Description | URL | Ref |
|---|---|---|---|
| ScÅtter | SAXS data reduction and analysis ($R_g$ and $P(r)$ determination, etc.) | http://www.bioisis.net | |
| Scatter | To calculates form and structure factors (1D and 2D), particularly useful for calculations of structure factors for different symmetry mesophases | https://www.esrf.eu/UsersAndScience/Experiments/CRG/BM26/SaxsWaxs/DataAnalysis/Scatter | [103] |
| SCT | Low resolution bead models of solution SAXS data, addition of hydration layers, etc. | http://dww100.github.io/sct | |
| SITUS | Calculates low resolution bead models of protein structures derived from SAXS data, for protein docking studies | https://situs.biomachina.org/tutorial_saxs.html | [104] |
| SLD Calculator | Calculates neutron and x-ray scattering length density neutron and coherent and incoherent cross-sections and penetration depth | https://sld-calculator.appspot.com/save | [105] |
| STOCHfit | Stochastic fitting of reflectivity data using dynamical theory for a slab model | http://stochfit.sourceforge.net | |
| WAXSiS | Calculates SAXS/WAXS profile for biomolecules in solution, based on explicit-solvent all-atom molecular dynamics (MD) simulations | http://waxsis.uni-goettingen.de/about | [106] |
| WillItFit | Form factor fitting. Does not come as an executable so not straightforward to install | https://sourceforge.net/projects/willitfit | [107] |
| X+ | Solution SAXS data reduction, analysis, and modelling | https://scripts.iucr.org/cgi-bin/paper?aj5158 | [108] |
| XRDUA | Detector file processing, can handle various SAXS file types (designed for powder diffraction data processing) | http://xrdua.ua.ac.be/index.php?id=xrduaoverview | |

Includes information from the list available at Ref. [109]. Does not include software specific to particular beamlines.

fits. The discussion in Section 1.7.4 about polydispersity effects potentially masking details of form factor features, should also be noted. It can sometimes be difficult to uniquely analyse form factors for slightly anisotropic structures for this reason. Instrumental smearing effects (Section 5.5.1) also wash out form factor features, and this can be problematic for detailed analysis of SANS form factor data. Consideration of external constraints based on other measurements or molecular parameters can address some of these issues.

## 2.18   SAS SOFTWARE

A wide variety of software is available for SAS data analysis, including dedicated software for data reduction and modelling. Table 2.2 provides an extensive list of software available to download or in some cases (for specific data analyses or modelling) on web servers. Much of this is freely available, although where this is not the case it is indicated in the table. As evident from an inspection of Table 2.2, some of this software is specific to a particular application, for instance GISAXS data reduction, or biomolecular solution SAXS data modelling.

## REFERENCES

1. Hamley, I.W. (2020). *Introduction to Peptide Science*. Chichester: Wiley.
2. Richards, E.G. (1968). A simplified method for the determination of the nucleotide composition of polyribonucleotides by spectrophotometric analysis. *European Journal of Biochemistry* 4: 256–264.
3. Graewert, M.A. and Jeffries, C.M. (2017). Sample and buffer preparation for SAXS. In: *Biological Small Angle Scattering: Techniques, Strategies and Tips. Advances in Experimental Medicine and Biology* (ed. B. Chaudhuri), 11–30. Berlin: Springer.
4. Lattman, E.E., Grant, T.D., and Snell, E.H. (2018). *Biological Small-Angle Scattering: Theory and Practice*. Oxford: Oxford University Press.
5. Pauw, B.R. (2013). Everything SAXS: small-angle scattering pattern collection and correction. *Journal of Physics. Condensed Matter* 25: 24.
6. Ghosh, R.E., Egelhaaf, S.U., and Rennie, A.R. (2012). *A Computing Guide for Small-Angle Scattering Experiments*. Grenoble: Institut Laue-Langevin.
7. Lindner, P. (2002). Scattering experiments. In: *Neutrons, X-Rays and Light Scattering Methods Applied to Soft Condensed Matter* (eds. P. Lindner and T. Zemb). Amsterdam: Elsevier.
8. Svergun, D.I., Koch, M. H. J., Timmins, P.A., and May, R.P. (2013). *Small Angle X-ray and Neutron Scattering from Solutions of Biological Macromolecules*. Oxford: Oxford University Press.
9. Narayanan, T., Diat, O., and Bosecke, P. (2001). SAXS and USAXS on the high brilliance beamline at the ESRF. *Nuclear Instruments and Methods in Physics Research Section A: Accelerators, Spectrometers, Detectors and Associated Equipment* 467: 1005–1009.

10. Narayanan, T. (2008). Synchrotron small-angle x-ray scattering. In: *Soft Matter Characterization* (eds. R. Borsali and R. Pecora), 899–952. Berlin: Springer-Verlag.

11. Brennich, M., Pernot, P., and Round, A. (2017). How to analyze and present SAS data for publication. In: *Biological Small Angle Scattering: Techniques, Strategies and Tips. Advances in Experimental Medicine and Biology* (ed. B. Chaudhuri), 47–64. Berlin: Springer.

12. Franke, D., Jeffries, C.M., and Svergun, D.I. (2015). Correlation map, a goodness-of-fit test for one-dimensional x-ray scattering spectra. *Nature Methods* 12: 419–422.

13. Nielsen, S.S., Toft, K.N., Snakenborg, D. et al. (2009). BioXTAS RAW, a software program for high-throughput automated small-angle x-ray scattering data reduction and preliminary analysis. *Journal of Applied Crystallography* 42: 959–964.

14. Hamley, I.W., Castelletto, V., Moulton, C.M. et al. (2010). Self-assembly of a modified amyloid peptide fragment: pH responsiveness and Nematic phase formation. *Macromolecular Bioscience* 10: 40–48.

15. Leadbetter, A.J. and Norris, E.K. (1979). Distribution functions in three liquid crystals from x-ray diffraction measurements. *Molecular Physics* 38: 669–686.

16. Deutsch, M. (1991). Orientational order determination in liquid crystals by x-ray diffraction. *Physical Review A* 44: 8264–8270.

17. Hamley, I.W. (1991). Scattering from uniform, cylindrically symmetric particles in liquid crystal phases. *Journal of Chemical Physics* 95: 9376–9383.

18. Stribeck, N. (2007). *X-ray Scattering of Soft Matter*. Berlin: Springer-Verlag.

19. Muthig, M., Prevost, S., Orglmeister, R., and Gradzielski, M. (2013). SASET: a program for series analysis of small-angle scattering data. *Journal of Applied Crystallography* 46: 1187–1195.

20. Roe, R.-J. (2000). *Methods of X-ray and Neutron Scattering in Polymer Science*. New York: Oxford University Press.

21. Hamley, I.W. (2001). Liquid crystals. In: *Encyclopedia of Chemical Physics and Physical Chemistry* (eds. J.H. Moore and N.D. Spencer), 2259–2284. Bristol: Institute of Physics.

22. Hamley, I.W. (2007). *Introduction to Soft Matter*. Revised Edition. Chichester: Wiley.

23. ESRF. (2020). Rat tail positions. The European Syncrotron. http://www.esrf.eu/UsersAndScience/Experiments/CRG/BM26/SaxsWaxs/Rattail

24. ESRF. (2020). NBS silicon powder. The European Syncrotron. http://www.esrf.eu/UsersAndScience/Experiments/CRG/BM26/SaxsWaxs/Silicon

25. ESRF. (2020). HDPE peak positions. The European Syncrotron. http://www.esrf.eu/UsersAndScience/Experiments/CRG/BM26/SaxsWaxs/Hdpe

26. Elettra-Sincrotrone Trieste. (2020). ESSAXS beamline. https://www.elettra.trieste.it/it/lightsources/elettra/elettra-beamlines/saxs/beamline/page-5.html?showall=

27. Dreiss, C.A., Jack, K.S., and Parker, A.P. (2006). On the absolute calibration of bench-top small-angle x-ray scattering instruments: a comparison of different standard methods. *Journal of Applied Crystallography* 39: 32–38.

28. NIST Center for Neutron Research. (2020). Neutron activation and scattering calculator. https://www.ncnr.nist.gov/resources/activation/

29. Glatter, O. (2018). *Scattering Methods and their Application in Colloid and Interface Science*, Chapter 5. Amsterdam: Elsevier.

30. Kell, G.S. (1975). Density, thermal Expansivity, and compressibility of liquid water from $0^\circ$ to $150\,^\circ C$ – correlations and tables for atmospheric-pressure and saturation reviewed and expressed an 1968 temperature scale. *Journal of Chemical & Engineering Data* 20: 97–105.

31. Wignall, G.D. (2011). Practical aspects of SANS experiments. In: *Neutrons in Soft Matter* (eds. T. Imae, T. Kanaya, M. Furusaka and N. Torikai), 285–310. Wiley Online Library: Wiley.

32. Wignall, G.D. (1991). Instrumental resolution effects in small-angle scattering. *Journal of Applied Crystallography* 24: 479–484.

33. Pedersen, J.S. and Riekel, C. (1991). Resolution function and flux at the sample for small-angle X-ray-scattering calculated in position- angle-wavelength space. *Journal of Applied Crystallography* 24: 893–909.
34. Pedersen, J.S., Posselt, D., and Mortensen, K. (1990). *Journal of Applied Crystallography* 23: 321.
35. Pedersen, J.S. (2002). Instrumentation for small-angle x-ray and neutron scattering. In: *Neutrons, X-rays and Light Scattering Methods Applied to Soft Condensed Matter* (eds. P. Lindner and T. Zemb). Amsterdam: Elsevier.
36. Jeffries, C.M., Graewert, M.A., Svergun, D.I., and Blanchet, C.E. (2015). Limiting radiation damage for high-brilliance biological solution scattering: practical experience at the EMBL P12 beamline PETRAIII. *Journal of Synchrotron Radiation* 22: 273–279.
37. Kuwamoto, S., Akiyama, S., and Fujisawa, T. (2004). Radiation damage to a protein solution, detected by synchrotron x-ray small-angle scattering: dose-related considerations and suppression by cryoprotectants. *Journal of Synchrotron Radiation* 11: 462–468.
38. Grishaev, A. (2012). Sample preparation, data collection, and preliminary data analysis in biomolecular solution x-ray scattering. *Current Protocols in Protein Science* 70: 17.4.
39. Crosas, E., Castellvi, A., Crespo, I. et al. (2017). Uridine as a new scavenger for synchrotron-based structural biology techniques. *Journal of Synchrotron Radiation* 24: 53–62.
40. Meisburger, S.P., Warkentin, M., Chen, H.M. et al. (2013). Breaking the radiation damage limit with Cryo-SAXS. *Biophysical Journal* 104: 227–236.
41. Spalla, O. (2002). General theorems in small-angle scattering. In: *Neutrons, X-rays and Light Scattering Methods Applied to Soft Condensed Matter* (eds. P. Lindner and T. Zemb). Amsterdam: Elsevier.
42. Jacques, D.A. and Trewhella, J. (2010). Small-angle scattering for structural biology-expanding the frontier while avoiding the pitfalls. *Protein Science* 19: 642–657.
43. Putnam, C.D., Hammel, M., Hura, G.L., and Tainer, J.A. (2007). X-ray solution scattering (SAXS) combined with crystallography and computation: defining accurate macromolecular structures, conformations and assemblies in solution. *Quarterly Reviews of Biophysics* 40: 191–285.
44. Chaudhuri, B., Muñoz, I.G., Qian, S., and Urban, V.S. (eds.) (2017). *Biological Small Angle Scattering: Techniques, Strategies and Tips*. Singapore: Springer.
45. Porod, G. (1951). Die Röntgenkleinwinkelstreung von Dichtdepackten Kolloidalen Systemen. *Kolloid-Zeitschrift und Zeitschrift für Polymere* 124: 83–114.
46. Debye, P., Anderson, H.R., and Brumberger, H. (1957). Scattering by an inhomogeneous solid. 2. The correlation function and its application. *Journal of Applied Physics* 28: 679–683.
47. Glatter, O. and Kratky, O. (eds.) (1982). *Small Angle X-ray Scattering*. London: Academic.
48. Rambo, R.P. and Tainer, J.A. (2011). Characterizing flexible and intrinsically unstructured biological macromolecules by SAS using the Porod-Debye law. *Biopolymers* 95: 559–571.
49. Hammel, M. (2012). Validation of macromolecular flexibility in solution by small-angle x-ray scattering (SAXS). *European Biophysics Journal with Biophysics Letters* 41: 789–799.
50. Benoit, H. (1953). On the effect of branching and polydispersity on the angular distribution of light scattered by Gaussian coils. *Journal of Polymer Science* 11: 507–510.
51. Kratky, O. and Porod, G. (1949). Röntgenuntersuchung gelöster Fadenmoleküle. *Recueil des Travaux Chimiques des Pays-Bas* 68 (12): 1106–1122.
52. Benoit, H. and Doty, P. (1953). Light scattering from non-Gaussian chains. *Journal of Physical Chemistry* 57: 958–963.
53. Peterlin, A. (1953). Modele statistique des grosses molecules a chaines courtes. V. Diffusion de la lumiere. *Journal of Polymer Science* 10.

54. Sharp, P. and Bloomfield, V.A. (1968). Light scattering from wormlike chains with excluded volume effects. *Biopolymers* 6: 1201–1211.

55. des Cloizeaux, J. (1973). Form factor of an infinite Kratky-Porod chain. *Macromolecules* 6: 403–407.

56. Koyama, R. (1973). Light scattering of stiff polymer chains. *Journal of the Physical Society of Japan* 34: 1029–1038.

57. Kholodenko, A.L. (1993). Analytical calculation of the scattering function for polymers of arbitrary flexibility using the Dirac propagator. *Macromolecules* 26: 4179–4183.

58. Shen, C.-L., Fitzgerald, M.C., and Murphy, R.M. (1994). Effect of acid predissolution on fibril size and fibril flexibility of synthetic $\beta$-amyloid peptide. *Biophysical Journal* 67: 1238–1246.

59. Brûlet, A., Boué, F., and Cotton, J.P. (1996). About the experimental determination of the persistence length of wormlike chains of polystyrene. *Journal de Physique (France) II* 6: 885–891.

60. Pedersen, J.S. and Schurtenberger, P. (1996). Scattering functions of semiflexible polymers with and without excluded volume effects. *Macromolecules* 29: 7602–7612.

61. Yamakawa, H. and Fujii, M. (1974). Light-scattering from wormlike chains – determination of shift factor. *Macromolecules* 7: 649–654.

62. Yoshizaki, T. and Yamakawa, H. (1980). Scattering functions of wormlike and helical wormlike chains. *Macromolecules* 13: 1518–1525.

63. Huber, K. and Burchard, W. (1989). Scattering behavior of wormlike star macromolecules. *Macromolecules* 22: 3332–3336.

64. Shen, C.-L. and Murphy, R.M. (1995). Solvent effects on self-assembly of b-amyloid peptide. *Biophysical Journal* 69: 640–651.

65. Pallitto, M.M., Ghanta, J., Heinzelman, P. et al. (1999). Recognition sequence design for peptidyl modulators of $\beta$-amyloid aggregation and toxicity. *Biochemistry* 38: 3570–3578.

66. O'Reilly, J.M., Teegarden, D.M., and Wignall, G.D. (1985). Small-angle and intermediate-angle neutron-scattering from Stereoregular poly(methyl methacrylate). *Macromolecules* 18: 2747–2752.

67. Bates, F. (1988). Small-angle neutron-scattering from amorphous polymers. *Journal of Applied Crystallography* 21: 681–691.

68. Kratky, O. (1963). X-ray small angle scattering with substances of biological interest in diluted solutions. *Progress in Biophysics and Molecular Biology* 13: 105–173.

69. Knoll, W., Haas, J., Stuhrmann, H.B. et al. (1981). Small-angle neutron-scattering of aqueous dispersions of lipids and lipid mixtures – a contrast variation study. *Journal of Applied Crystallography* 14: 191–202.

70. Higgins, J.S. and Benoît, H.C. (1994). *Polymers and Neutron Scattering*. Oxford: Oxford University Press.

71. Hadziioannou, G. and Stein, R.S. (1984). Neutron-scattering studies of dimensions and of interactions between components in polystyrene poly(vinyl methyl-ether) and poly(vinylidene fluoride) poly(methyl methacrylate) amorphous blends. *Macromolecules* 17: 567–573.

72. Hamley, I.W., Fairclough, J.P.A., King, S.M. et al. (1997). A small angle neutron scattering study of the conformation of a side chain liquid crystal poly(methacrylate) in the smectic C phase. *Liquid Crystals* 22: 679–684.

73. Feigin, L.A. and Svergun, D.I. (1987). *Structure Analysis by Small-Angle X-ray and Neutron Scattering*. New York: Plenum.

74. Pedersen, J.S. (1997). Analysis of small-angle scattering data from colloids and polymer solutions: modeling and least-squares fitting. *Advances in Colloid and Interface Science* 70: 171–210.

75. Pedersen, J.S. (2002). Modelling of small-angle scattering data. In: *Neutrons, X-rays and Light Scattering Methods Applied to Soft Condensed Matter* (eds. P. Lindner and T. Zemb). Amsterdam: Elsevier.

76. Poitevin, F., Orland, H., Doniach, S. et al. AquaSAXS: a web server for computation andfitting of SAXS profiles with non-uniformallyhydrated atomic models. *Nucleic Acids Research* 39: W184–W189.

77. Franke, D., Petoukhov, M.V., Konarev, P.V. et al. (2017). ATSAS 2.8: a comprehensive data analysis suite for small-angle scattering from macromolecular solutions. *Journal of Applied Crystallography* 50: 1212–1225.

78. Grishaev, A., Guo, L.A., Irving, T., and Bax, A. (2010). Improved fitting of solution x-ray scattering data to macromolecular structures and structural ensembles by explicit water modeling. *Journal of the American Chemical Society* 132: 15484–15486.

79. Hansen, S. (2012). BayesApp: a web site for indirect transformation of small-angle scattering data. *Journal of Applied Crystallsography* 45: 566–567.

80. Pelikan, M., Hura, G.L., and Hammel, M. (2009). Structure and flexibility within proteins as identified through small angle x-ray scattering. *General Physiology and Biophysics* 28: 174–189.

81. Pospelov, G., Van Herck, W., Burle, J. et al. (2020). BornAgain: software for simulating and fitting grazing-incidence small-angle scattering. *Journal of Applied Crystallography* 53: 262–276.

82. Benecke, G., Wagermaier, W., Li, C.H. et al. (2014). A customizable software for fast reduction and analysis of large x-ray scattering data sets: applications of the new DPDAK package to small-angle x-ray scattering and grazing-incidence small-angle x-ray scattering. *Journal of Applied Crystallography* 47: 1797–1803.

83. Hammersley, A.P. (2016). FIT2D: a multi-purpose data reduction, analysis and visualization program. *Journal of Applied Crystallography* 49: 646–652.

84. Babonneau, D. (2010). FitGISAXS: software package for modelling and analysis of GISAXS data using IGOR pro. *Journal of Applied Crystallography* 43: 929–936.

85. Schneidman-Duhovny, D., Hammel, M., Tainer, J.A., and Sali, A. (2013). Accurate SAXS profile computation and its assessment by contrast variation experiments. *Biophysical Journal* 105: 962–974.

86. Schneidman-Duhovny, D., Hammel, M., Tainer, J.A., and Sali, A. (2016). FoXS, FoXS-Dock and MultiFoXS: single-state and multi-state structural modeling of proteins and their complexes based on SAXS profiles. *Nucleic Acids Research* 44: W424–W429.

87. Spinozzi, F., Ferrero, C., Ortore, M.G. et al. (2014). GENFIT: software for the analysis of small-angle x-ray and neutron scattering data of macromolecules in solution. *Journal of Applied Crystallography* 47: 1132–1139.

88. Fritz, G., Bergmann, A., and Glatter, O. (2000). Evaluation of small-angle scattering data of charged particles using the generalized indirect Fourier transformation technique. *Journal of Chemical Physics* 113: 9733–9740.

89. Chourou, S.T., Sarje, A., Li, X.Y.S. et al. (2013). HipGISAXS: a high-performance computing code for simulating grazing-incidence x-ray scattering data. *Journal of Applied Crystallography* 46: 1781–1795.

90. Ilavsky, J. and Jemian, P.R. (2009). Irena: tool suite for modeling and analysis of small-angle scattering. *Journal of Applied Crystallography* 42: 347–353.

91. Lazzari, R. (2002). IsGISAXS: a program for grazing-incidence small-angle x-ray scattering analysis of supported islands. *Journal of Applied Crystallography* 35: 406–421.

92. Arnold, O., Bilheux, J.C., Borreguero, J.M. et al. (2014). Mantid-data analysis and visualization package for neutron scattering and mu SR experiments. *Nuclear Instruments and Methods in Physics Research Section A: Accelerators, Spectrometers, Detectors and Associated Equipment* 764: 156–166.

93. Pauw, B.R., Pedersen, J.S., Tardif, S. et al. (2013). Improvements and considerations for size distribution retrieval from small-angle scattering data by Monte Carlo methods. *Journal of Applied Crystallography* 46: 365–371.

94. Whitten, A.E., Cai, S.Z., and Trewhella, J. (2008). MULCh: modules for the analysis of small-angle neutron contrast variation data from biomolecular assemblies. *Journal of Applied Crystallography* 41: 222–226.

95. Tate, M.P., Urade, V.N., Kowalski, J.D. et al. (2006). Simulation and interpretation of 2D diffraction patterns from self-assembled nanostructured films at arbitrary angles of incidence: from grazing incidence (above the critical angle) to transmission perpendicular to the substrate. *Journal of Physical Chemistry B* 110: 9882–9892.

96. Kline, S.R. (2006). Reduction and analysis of SANS and USANS data using IGOR pro. *Journal of Applied Crystallography* 39: 895–900.

97. Ilavsky, J. (2012). Nika: software for two-dimensional data reduction. *Journal of Applied Crystallography* 45: 324–328.

98. Maranville, B.B. (2017). Interactive, web-based calculator of neutron and X-ray reflectivity. *Journal of Research of the National Institute of Standards and Technology* 122: 6.

99. Bressler, I., Kohlbrecher, J., and Thünemann, A.F. (2015). SASfit: a tool for small-angle scattering data analysis using a library of analytical expressions. *Journal of Applied Crystallography* 48: 1587–1598.

100. Curtis, J.E., Raghunandan, S., Nanda, H., and Krueger, S. (2012). SASSIE: a program to study intrinsically disordered biological molecules and macromolecular ensembles using experimental scattering restraints. *Computer Physics Communications* 183: 382–389.

101. Liu, H.G., Hexemer, A., and Zwart, P.H. (2012). The small angle scattering ToolBox (SASTBX): an open-source software for biomolecular small-angle scattering. *Journal of Applied Crystallography* 45: 587–593.

102. Piiadov, V., de Araujo, E.A., Neto, M.O. et al. (2019). SAXSMoW 2.0: online calculator of the molecular weight of proteins in dilute solution from experimental SAXS data measured on a relative scale. *Protein Science* 28: 454–463.

103. Forster, S., Apostol, L., and Bras, W. (2010). Scatter: software for the analysis of nano- and mesoscale small-angle scattering. *Journal of Applied Crystallography* 43: 639–646.

104. Wriggers, W. and Chacon, P. (2001). Using Situs for the registration of protein structures with low-resolution bead models from X-ray solution scattering. *Journal of Applied Crystallography* 34: 773–776.

105. Danauskas, S.M., Li, D.X., Meron, M. et al. (2008). Stochastic fitting of specular X-ray reflectivity data using StochFit. *Journal of Applied Crystallography* 41: 1187–1193.

106. Knight, C.J. and Hub, J.S. (2015). WAXSiS: a web server for the calculation of SAXS/WAXS curves based on explicit-solvent molecular dynamics. *Nucleic Acids Research* 43: W225–W230.

107. Pedersen, M.C., Arleth, L., and Mortensen, K. (2013). WillItFit: a framework for fitting of constrained models to small-angle scattering data. *Journal of Applied Crystallography* 46: 1894–1898.

108. Ben-Nun, T., Ginsburg, A., Szekely, P., and Raviv, U. (2010). X plus: a comprehensive computationally accelerated structure analysis tool for solution X-ray scattering from supramolecular self-assemblies. *Journal of Applied Crystallography* 43: 1522–1531.

109. SAS Portal. (2020). http://smallangle.org/content/software

# 3

# Instrumentation for SAXS and SANS

## 3.1 INTRODUCTION

This chapter discusses the instrumentation used at typical small-angle x-ray scattering (SAXS) and small-angle neutron scattering (SANS) beamlines, covering in outline the design of the beamlines along with the essential optical components, and also reviewing sample environments including both standard sample holders and selected specialist cells for in situ measurements under different conditions.

SAXS measurements have the advantage that they can be performed on laboratory instruments as well as synchrotron facilities and this chapter discusses both types of instrument. Neutron scattering experiments require a neutron source, of which there are two types – nuclear reactors and spallation sources, as outlined in this chapter. The setup of SANS beamlines is discussed, as is the typical configuration of ultra-small-angle neutron scattering (USANS) and ultra-small-angle x-ray scattering (USAXS) instruments.

This chapter is organized as follows. Section 3.2 lists major currently available international synchrotron facilities, while Section 3.3 similarly summarizes global SANS facilities at both reactor and spallation sources. The following sections are dedicated to instrument setups, Section 3.4 dealing with synchrotron SAXS instrumentation, and Section 3.5 with laboratory SAXS equipment. Instrumentation at SANS beamlines is described in Section 3.6, while Section 3.7 considers equipment for USAXS and USANS measurements, typically a Bonse-Hart camera. The following sections

*Small-Angle Scattering: Theory, Instrumentation, Data and Applications*,
First Edition. Ian W. Hamley.
© 2021 John Wiley & Sons Ltd. Published 2021 by John Wiley & Sons Ltd.

discuss sample environments. Standard sample environments for static and variable temperature SAXS measurements are the focus of Section 3.8, while Sections 3.9 and 3.10, respectively, outline standard sample cells for SANS and GISAS experiments. Microfocus or microbeam SAXS (and WAXS or x-ray diffraction), which permit investigation of the order in a sample scanning a focussed beam across it, are discussed in Section 3.11. Section 3.12 then discusses a range of nonstandard sample environments that are increasingly used during kinetic measurements after stopped flow, temperature changes, and for in situ measurements to probe the effects of different types of field (mechanical, electrical, magnetic) on materials, among others.

## 3.2  SYNCHROTRON FACILITIES

SAXS beamlines are available at many synchrotron facilities around the world. Figure 3.1 shows third-generation synchrotron facilities internationally. There are several other smaller and previous generation facilities around the world.

Table 3.1 summarizes synchrotron facilities around the world, along with an estimate of currently available SAXS beamlines, with some remarks on which type of SAXS beamline is operational.

## 3.3  NEUTRON SCATTERING FACILITIES

Neutron scattering facilities are scattered (!) around the developed world, as shown in Figure 3.2, which provides a map of the main international

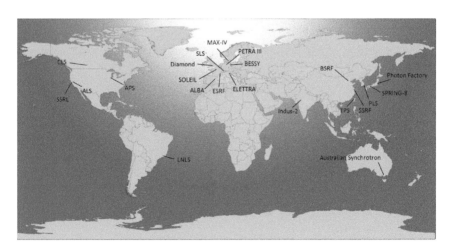

**Figure 3.1**  Major international third-generation synchrotron facilities.

Table 3.1 International synchrotron scattering facilities.

| Name | Beamlines (SAXS beamlines) | Remarks |
| --- | --- | --- |
| APS: Advanced Photon Source (Argonne, USA) | 71 (7) | Several of the beamlines are multipurpose but can do SAXS and some SAXS/WAXS or GISAXS or XPCS. One beamline SAXS with fibre diffraction. One capable of USAXS |
| ALS: Advanced Light Source (Berkeley, USA) | 40 (2) | One SAXS/WAXS/GISAXS and one BioSAXS (with MX) beamline |
| SSRL: Stanford Synchrotron Research Laboratory (Stanford, USA) | 30 (2) | One materials SAXS (and SAXS/WAXS) and one BioSAXS beamline |
| CHESS: Cornell High Energy Synchrotron Source | 7 (2) | One SAXS/WAXS and one BioSAXS beamline |
| CLS: Canadian Light Source (Saskatoon, Canada) | 16 (0) | One beamline able to do GISAXS and another to do USAXS |
| LNLS: Laboratório Nacional de Luz Síncrotron (Campinas, Brazil) | 17 (2) | WAXS possible on both SAXS beamlines |
| Diamond Light Source (Harwell, UK) | 31 (2) | One BioSAXS beamline and one materials science SAXS beamline |
| Synchrotron SOLEIL (Gif-sur-Yvette, France) | 29 (1) | One SAXS/WAXS beamline |
| ALBA (Cerdanyola del Vallès, Spain) | 9 (1) | One SAXS/WAXS beamline |
| ESRF: European Synchrotron Radiation Facility (Grenoble, France) | 49 (3) | Three main SAXS beamline – one general purpose, one SAXS/WAXS, one BioSAXS. Other beamlines able to do GISAXS. |

Table 3.1  (Continued)

| Name | Beamlines (SAXS beamlines) | Remarks |
| --- | --- | --- |
| SLS: Swiss Light Source (Villigen, Switzerland) | 22 (1) | Flexible beamline able to do XPCS, nanotomography also |
| PETRA III at DESY (Hamburg, Germany) | 21 (2) | One SAXS/WAXS and one BioSAXS beamline |
| MAX-IV (Lund, Sweden) | 8 | New facility, other beamlines being commissioned including a SAXS beamline |
| BESSY: Berliner Elektronenspeicherring-Gesellschaft für Synchrotronstrahlung (Berlin, Germany) | 38 (1) | SAXS possible on a (N)EXAFS/XRF beamline |
| ELETTRA (Trieste, Italy) | 28 (1) | GISAXS and WAXS possible |
| Indus-2 (Indore, India) | 14 (1) | SAXS/WAXS beamline |
| BSRF: Beijing Synchrotron Radiation Facility (Beijing, China) | 14 (1) | SAXS/WAXS beamline |
| SSRF: Shanghai Synchrotron Radiation Facility (Shanghai, China) | 14 (1) | SAXS/WAXS beamline |
| TPS: Taiwan Photon Source (Hsinchu, Taiwan) | 7 (0) | Other beamlines being commissioned including SAXS beamlines |
| PLS: Pohang Light Source (Pohang, S. Korea) | 32 (3) | Two SAXS and one USAXS beamline |
| SPRING-8 (Hyogo, Japan) | 57 (3) | One SAXS/WAXS beamline, one BioSAXS beamline (not general access) and one industrial SAXS beamline. Another two beamlines capable of USAXS |
| Photon Factory (Tsukuba, Japan) | 48 (3) | SAXS/WAXS beamlines, also GISAXS |
| Siam Photon Source, SLRI (Korat, Thailand) | 12 (1) | SAXS/WAXS beamline |
| Australian Synchrotron (Clayton, Australia) | 8 (1) | SAXS/WAXS, GISAXS and ASAXS possible |

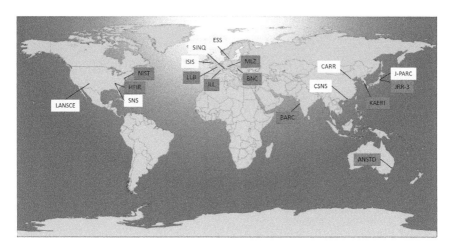

**Figure 3.2** International neutron scattering facilities – reactors shown by red squares, spallation sources by yellow squares.

laboratories (there are also a number of smaller research reactors with limited number of beamlines and/or user programmes). These facilities are divided into those in which neutrons are sourced from nuclear reactors and from spallation sources. These are discussed in turn below and are listed in Table 3.2.

## 3.4 SYNCHROTRON SAXS INSTRUMENTATION

A synchrotron consists of an accelerator ring in which electrons are accelerated to relativistic speeds (i.e. a high fraction of the speed of light), under which conditions they emit electromagnetic radiation (Cherenkov radiation). This spans the range from infrared radiation to x-rays. For SAXS experiments, hard x-rays are employed with a typical wavelength $\lambda = 1$ Å. Figure 3.3 shows a picture of a synchrotron along with a representative plan of a beamline. A beamline comprises the x-ray optics (enclosed and not usually accessed by users), the experimental hutch containing the sample environment and the SAXS camera and detector(s), and a separate control room. For obvious safety reasons, researchers cannot be present in the experimental hutch when a synchrotron x-ray beam is on. The hutch is controlled by an interlock system.

Figure 3.4 shows a schematic of a synchrotron, along with configurations of magnets used to steer and focus the beam. The word *synchrotron* derives from the *synchronization* of the variation of the magnetic fields used to confine the electrons with the energy of the accelerating electrons. A linear

Table 3.2   International neutron scattering facilities.

| Name | Beamlines (SANS beamlines) | Remarks |
|---|---|---|
| LANSCE: Los Alamos Neutron Science Center (Los Alamos, USA) | 5 (0) | Does not currently have a dedicated SANS beamline |
| HFIR: High Flux Isotope Reactor (Oak Ridge National Lab, USA) | 13 (2) | One biological and one 'general purpose' SANS instrument |
| SNS: Spallation Neutron Source (Oak Ridge National Lab, USA) | 20 (2) | One 'extended q-range' SANS beamline and one USANS beamline based on Bonse–Hart camera |
| NIST: National Institute of Standards and Technology Institute for Neutron Research (Gaithersburg, USA) | 28 (5) | One 10 m instrument, two 30 m instruments, one 45 m instrument (capable of very low $q$ measurements to $10^{-4}$ Å$^{-1}$) and one USANS instrument |
| ISIS Neutron and Muon Source (Harwell, UK) | 36 (4) | General purpose SANS beamlines including high flux beamlines, plus polarized SANS and spin echo SANS capabilities |
| Reactor Institute Delft | 6 (2) | SANS and SESANS |
| LLB: Laboratoire Léon Brillouin (Saclay, France) | 23 (4) | Closed in 2019 |
| ILL: Institut Laue–Langevin (Grenoble, France) | 48 (4) | Three main SANS diffractometers include wide (and dynamic) $q$ range |
| SINQ: Swiss Spallation Neutron Source (Villigen, Switzerland) | 22 (1) | One SANS beamline (another planned) and one USANS beamline |
| ESS: European Spallation Source | In construction | |
| MLZ: Heinz Maier-Leibnitz Zentrum (Garching, Germany), also known as FRM-II Forschungsreacktor München II | 35 (4) | Three SANS instruments and one very small angle instrument ($q$ down to $4 \times 10^{-5}$ Å$^{-1}$) |

| | | |
|---|---|---|
| BNC: Budapest Neutron Centre (Budapest, Hungary) | 18 (2) | One SANS instrument and one capable of very small angle measurements ($q$ down to $3 \times 10^{-4}$ Å$^{-1}$) |
| BARC: Dhruva reactor at Bhabhi Atomic Research Centre (Mumbai, India) | 13 (2) | |
| CARR: China Advanced Research Reactor (Beijing, China) | Commissioning | |
| CSNS: China Spallation Neutron Source (Guangdong, China) | 3 (1) others commissioning | |
| KAERI: Korea Atomic Energy Research Institute (Daejeon, Korea) | 15 (3) | Two SANS and one USANS beamline |
| J-PARC (Tokai, Japan) | 21 (1) | |
| JRR-3: Japan Atomic Energy Agency Research Reactor 3 (Tokai, Japan) | 31 (2) | One SANS and one USANS 15 user access beamlines |
| ANSTO | 15 (3) | Two SANS and one USANS beamline (SANS capable) |

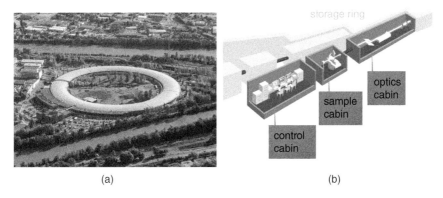

(a)                                                    (b)

**Figure 3.3** (a) Aerial photograph of the European Synchrotron Radiation Facility, (b) Schematic of a typical synchrotron beamline.

accelerator (LINAC) generates an electron beam, the energy of which is increased in a booster ring before injection into the storage ring where the electrons circulate at fixed energy for a lifetime (at most synchrotrons) of 12–24 hours, with top-up injections after this period (or complete refilling if beam is lost, after shutdowns, etc.). In the storage ring, the beam of electrons is steered by either bending magnets or insertion devices, which are *wigglers* (multipole wigglers) or *undulators*. These feed the x-rays into beamlines located in sectors of the synchrotron ring. Synchrotron rings have a circumference of hundreds of meters (the current world record is the 2.3 km circumference PETRA III synchrotron ring at DESY in Germany [1]). The radius of curvature of the electron beam, $R_c$ (in meters) is related to the electron energy and magnetic field strength by $R_c = 3.335E(\text{GeV})/B(\text{T})$ [2]. Insertion devices are arrays of magnetic dipoles that oscillate the electron beam in the horizontal plane. Wigglers create larger oscillations and produce a broader spectrum of radiation than undulators, as well as an asymmetric beam profile (Figure 3.4). Undulators produce the most intense radiation, at defined harmonics, which can be tuned to desired energies/wavelengths on beamlines such as SAXS beamlines. The flux increases as a function of the number of dipoles in the undulator. Further information on bending magnets and insertion devices is available elsewhere [3]. The radiofrequency cavity system in the synchrotron ring boosts the energy and maintains the circular orbit of the electrons (Figure 3.4).

Figure 3.5 shows a representative beamline diagram for the American Light Source, as an example to show the number of bending magnet and insertion device beamlines available. The major synchrotron facilities in Europe, Asia, and North America have similar numbers and distributions of beamlines (see Table 3.1, for example).

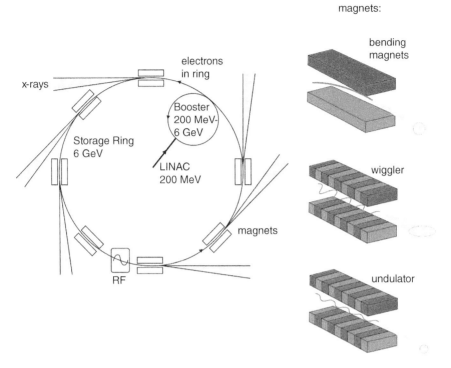

**Figure 3.4** Schematic of a synchrotron and magnet configurations (opposite dipoles are coloured red and blue, the green line is an electron beam and yellow indicates synchrotron light, i.e. photons). The energies are those of the ESRF (European Synchrotron Radiation Facility, Grenoble, France). RF indicates radiofrequency cavity.

Synchrotrons typically operate with a current of 100–500 mA (lower in low-bunch modes). Synchrotrons produce pulsed beams of x-rays. The x-rays are bunched with a typical bunch duration of 40 ps–20 ns, with a gap of 2 ns–1 μs between bunches. The bunches may be injected uniformly or distributed in multi-bunch modes (see e.g. [5]) or single bunch or hybrid mode, the latter being used for specific time-resolved studies. In addition, x-rays produced by synchrotrons are highly polarized horizontally, with an additional component vertically polarized. This property is almost never used in small-angle scattering studies. The polarization factor is discussed in Section 4.3.

There are many variations of synchrotron SAXS beamline designs, although there are common features. To illustrate the sequence of optical elements, Figure 3.6 shows a schematic of a typical SAXS beamline, which has additional WAXS capability.

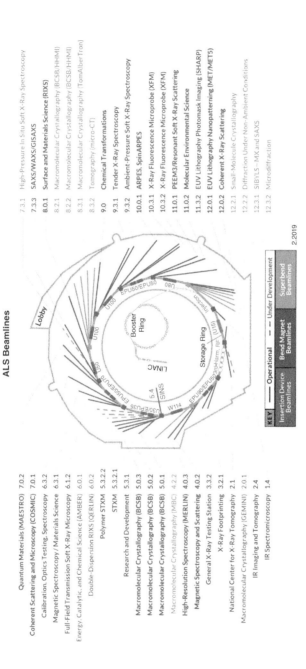

**Figure 3.5** Beamline plan for the American Light Source synchrotron. Superbend beamlines have superconducting bending magnets which increases x-ray flux and brightness. *Source*: From Ref. [4]. © 2010 The Regents of the University of California, through the Lawrence Berkeley National Laboratory.

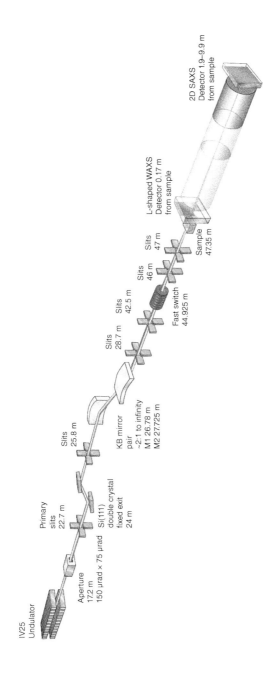

**Figure 3.6** Typical optics at a synchrotron SAXS/WAXS beamline (I22, Diamond). KB mirror denotes Kirkpatrick–Baez mirrors. The distances are with respect to the source (here an undulator). This beamline incorporates a fast switch as an optional element in the beam path. *Source*: From Ref. [6]. © 2019, Diamond Light Source.

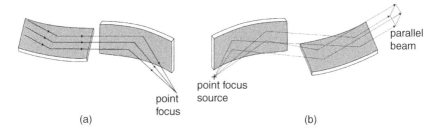

**Figure 3.7**    Schematics of (a) Kirkpatrick-Baez focussing mirrors, (b) Göbel multi-layer cross-coupled mirror collimation.

The synchrotron beam is usually monochromated using a Si(111) double-crystal monochromator; slits define the beam before focussing using Kirkpatrick-Baez curved monochromators (Figure 3.7a). Further slits define the beam before it is scattered by the sample. Evacuated tubes are essential to reduce air scattering between the sample and detector. On some beamlines, the detector is contained on a rail within a large, evacuated tank. On SAXS/WAXS beamlines, the WAXS detector is placed close to the sample, usually offset above it.

Most modern synchrotron SAXS beamlines use HPC (hybrid pixel counting) detectors, known as hybrid photon counting detectors (HPCDs) in single photon counting mode. These are semiconductor diode arrays. An incident photon produces many more electron-hole pairs in a solid-state semiconductor material than in a gas (gas multiwire detectors were widely used until recently and are still employed in some laboratory instruments) and so HPC arrays are highly efficient detectors [7]. Each pixel is addressed separately, achieving high count rates ($10^8$ photons $s^{-1}$ $mm^{-2}$) and photon counting sensitivity. Modern microfabrication methods enable the production of arrays with pixel sizes down to $75\,\mu m$ in a $4150 \times 4371$ array.

Previous types of detectors included multiwire gas-filled position-sensitive detectors (1D and 2D versions), charge-coupled device (CCD) cameras, proportional counters, and image plates. These technologies are reviewed elsewhere [8, 9]. CCD detectors are still in use in several synchrotron beamlines due to their high spatial sensitivity and also as WAXS detectors since the requirement for WAXS detectors to be close to the sample means that they cannot be too large.

A disadvantage of HPC detectors, which is especially notable for oriented samples, is that these detectors are built from microstrip arrays with dead zones in between producing arrays of stripes in the raw 2D detector images (Figure 2.3).

## 3.5   LABORATORY SAXS INSTRUMENTATION

Table 3.3 summarizes the configuration of the main currently marketed lab SAXS instruments.

The principles behind the production of x-rays are discussed in Section 4.2. Sealed tube x-ray sources are the traditional source, largely superseded in modern SAXS instruments by alternative sources with higher brilliance. Rotating anode sources comprising rotating solid metal discs were popular some decades ago but are complicated to maintain due to the extreme conditions since the temperature of the anode can reach 1000 °C and the anode is rotated at thousands of revolutions per minute. Microfocus x-ray sources are far easier to maintain and also produce beams focussed to 50 µm or less. They can achieve high brilliance (Eq. (4.1)) despite low power requirements. Microfocus sources of x-rays were incorporated into lab SAXS instruments in the 2000s, providing a better-defined beam and a high flux at the sample. Recently, liquid metal jet technology has been introduced on state-of-the-art instruments. This technology uses a microfocus x-ray tube that contains a jet of liquid metal (indium or gallium alloys are used), providing heat dissipation at high power loads. This produces a highly focussed beam with enhanced brightness compared to rotating anode sources.

A variety of collimation systems are employed, depending on the SAXS camera geometry and source. Figure 3.8 shows some arrangements. The classical Kratky camera produces a line-focussed beam (about 10 mm horizontally), and hence the scattering data are smeared by the beam profile. However, some commercial Kratky cameras now have two-dimensional focussing using multilayer optics. Pinhole collimation is a traditional method; scatterless slit systems have been adopted more recently. Finally, pairs of orthogonal parabolic mirrors (Göbel mirrors) produce a horizontal and vertically parallel beam. Figure 3.7b shows the principle by which a parabolic-focussing Göbel mirror produces a parallel beam. Note that Göbel mirrors have a different shape (parabolic) than Kirkpatrick-Baez (KB) mirrors used to focus x-rays (Figure 3.7a).

The usual pinhole collimation lab SAXS configurations also include a curved quartz crystal to monochromate and focus the beam horizontally and a glass mirror to provide vertical focus. This is the Huxley-Holmes camera [10, 11]. Franks optics, which are sometimes employed, consist of a pair of vertical and horizontal focussing mirrors that provide a point focus. In this case, a monochromating crystal is not used [10, 11].

As evident in Table 3.2, advanced lab SAXS instruments are now typically installed with hybrid photon counting (HPC) detectors due to

Table 3.3  Currently marketed lab SAXS instruments.

| Supplier | Source | Collimation | Detector | Capabilities (in addition to SAXS) |
|---|---|---|---|---|
| Anton Parr SAXSpace | Sealed tube | Kratky camera block beam collimation | Dectris1D Mythen2 R series or 2D EIGER R series HPC detectors | Simultaneous SAXS/WAXS Multiple sample environments including GISAXS stage |
| Anton Parr SAXSpoint | Microfocus source or liquid metal jet | Scatterless slits | Dectris EIGER R HPC detector for SAXS and another for high-resolution WAXS | Simultaneous SAXS/WAXS Multiple sample environments including GISAXS stage |
| Bruker Nanostar | Microfocus source or liquid metal jet | Parabolic multilayer Göbel mirrors (Montel optics) | VÄNTEC 2000 photon counting detector | Add-on system for WAXS and GISAXS stage |
| Rigaku NanoPix | High power point focus | Pinhole slits | Hypix-3000 HPC detector | Simultaneous SAXS/WAXS Multiple sample environments including GISAXS stage |
| Rigaku NanoPix Mini | Sealed tube | Bonse–Hart (line focus) geometry | Rigaku Ultrax 250 linear detector | Designed for solids |
| Rigaku BioSAXS 2000 | Microfocus source | Kratky camera, block beam collimation | Hypix-3000 HPC detector | Designed for BioSAXS |
| Rigaku NanoMax | Sealed tube or rotating anode | Kratky camera, block beam collimation | Can be fitted with a variety of detectors | Simultaneous SAXS/WAXS possible |
| Xenocs Xeuss 3 | Microfocus source or liquid metal jet | 2D multilayer mirror optics | Dectris Pilatus or Eiger HPC detectors | Simultaneous SAXS/WAXS Multiple sample environments including GISAXS stage |
| Xenocs BioXolver | Microfocus source or liquid metal jet | 2D multilayer mirror optics | Dectris HPC detector | Designed for BioSAXS with BioSAXS robot In-line SEC and UV–vis possible |

*Source:* Adapted from Ref. [2].

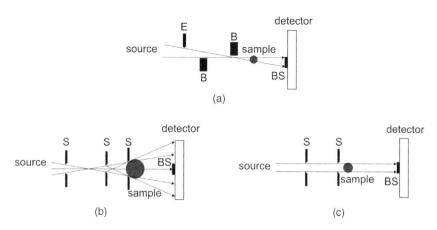

**Figure 3.8** SAXS collimation geometries. (a) Block collimation in a Kratky camera, (b) Pinhole collimation, (c) Slit collimation. *Sources*: Based on Svergun et al. [2] and Pedersen [10]. E: edge, B: block, BS: beam stop, S: slit.

their sensitivity and wide dynamic range. These have superseded previous detectors such as gas-filled multiwire detectors. HPC detectors are discussed in more detail in Section 3.4.

## 3.6 SANS INSTRUMENTATION

Figure 3.9 shows an example of a SANS beamline at a reactor source (D22 at the ILL, Grenoble, France). At a typical SANS beamline, as in a synchrotron beamline (Figure 3.6) the optics (velocity selector, collimation, focussing) are not accessible to the user and there is a separate experimental hutch for

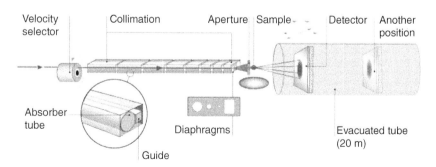

**Figure 3.9** Schematic configuration of SANS beamline D22 at the ILL. The neutrons are incident from the left. *Source*: From Ref. [12]. © 2019, ILL Neutrons for Society.

mounting samples. This is separated from the control room where the users sit to programme the control software and monitor the data. The fast neutrons produced by the reactor are first moderated using water or heavy water (slowed to typical thermal energies that produces neutrons with a wavelength $\lambda = 1.6–1.7$ Å [7, 8], which is too small for most purposes), and are then further moderated as described below for spallation beamlines (corresponding to a so-called cold source), to produce a peak in the range $\lambda = 4–6$ Å [7]. These slow neutrons are collimated using a mechanical velocity selector into a nondivergent beam, and then they pass through an aperture that defines the beam footprint on the sample. The scattered neutrons are then detected using a $^3$He detector on a moving rail in a large, evacuated tank.

The configuration of SANS beamlines at spallation sources is similar to that for a reactor. The moderation of the neutrons before entering the beamline is important due to the energy distribution of neutrons from a spallation source, extending to nonthermal neutrons. Moderation can be performed using a liquid $H_2$ moderator (at 20–25 K) or solid $CH_4$ (at 20–26 K) [8, 13, 14]. Background comprising low-wavelength neutrons and gamma rays is removed via velocity selectors, which may differ from those used for reactor beamlines. The velocity selector may be a Soller super-mirror bender comprising parallel vertical curved multilayers such as Ti/Ni [15]. These bend the neutron beam by reflection on the layered mirror material.

SANS performed at spallation sources relies on time-of-flight measurements, using a wide spectrum of the white beam of neutrons produced by these sources. This means that a beam with a distribution of wavelengths is incident on the sample. Using a pulsed beam, the time-of-flight of neutrons over a given path can be measured, and this provides the distribution of scattered neutrons as a function of $\lambda$ or $q$ via de Broglie's equation $\lambda = h/mv = ht/mL$ (where $m$ is the neutron mass, $m = 1.675 \times 10^{-27}$ kg) for time-of-flight $t$ over sample-detector distance $L$ [14, 16].

Neutrons at SANS beamlines have wavelengths in the typical range 2–20 Å [10, 14]. On reactor instruments, the flux at the sample is typically peaked towards the lower wavelength range. Thermal neutrons used in SANS instruments have speeds of $v \sim 1000$ ms$^{-1}$ (using the de Broglie relationship $v = h/m\lambda$, with a typical $\lambda = 4$ Å), which means that they can be separated according to speed (wavelength), i.e. this enables a monochromator to be based on mechanical methods. Figure 3.10 shows a mechanical velocity selector, which is a device resembling an aircraft turbine that consists of a series of metal blades coated with a strong neutron absorber such as boron (Table 2.1). The velocity selector rotates, producing a pulsed monochromated neutron beam, and it usually achieves a wavelength spread $\Delta\lambda/\lambda \sim 10\%$, which is typical for many SANS beamlines

**Figure 3.10** Mechanical neutron velocity selector.

at reactor sources. Only neutrons with a velocity within a certain range are able to pass between the rotating blades. Other designs have also been developed including monochromators comprising series of rotating discs with small slots. An alternative neutron beam monochromator system relies on reflection from pyrolytic graphite [13, 17].

Collimators are used to define the direction and divergence of a neutron beam. They are usually slits or pinholes in an absorbing material such as Gd or Cd foils or boron-loaded plastic [18, 19]. These need to be carefully spaced so that beam from the source does not reach the sample position.

On SANS beamlines, the aperture usually provides a beam of around 1 cm diameter at the sample. Focussing devices for neutrons such as compound refractive lenses or grazing incidence reflection optics have been proposed but not yet widely introduced [17].

Neutron detectors are gas-filled multiwire detectors containing $^3$He or arrays of tubes containing $^3$He. The isotope $^3$He is used to detect neutrons because it absorbs thermal neutrons producing $^1$H and $^3$H ions, which can then be detected electrically. To stop ion chain reactions, $CF_4$ or $CH_4$-Ar mixtures are used as stopping gases. SANS beamline neutron detectors usually comprise $64 \times 64$ or $128 \times 128$ pixels.

On reactor sources, a wide $q$ range is covered using different sample-detector distances (typically in the range 3–30 m) and wavelengths (3–20 Å). The SANS profiles are joined at overlapping regions (it is often noticeable that the resolution in the different ranges is different. The minimum and

**Table 3.4**  Available $q$ range assuming a detector with area $1\,m^2$ for different sample-to-detector distances and wavelengths [2].

| Sample-detector distance/m | Wavelength/nm | $q_{min}$/nm$^{-1}$ | $q_{max}$/nm$^{-1}$ |
|---|---|---|---|
| 1.0 | 0.5 | 0.628 | 7.556 |
|  | 1.0 | 0.314 | 3.778 |
| 2.0 | 0.5 | 0.314 | 4.211 |
|  | 1.0 | 0.157 | 2.106 |
| 5.0 | 0.5 | 0.126 | 1.747 |
|  | 1.0 | 0.063 | 0.873 |
| 10.0 | 0.5 | 0.063 | 0.878 |
|  | 1.0 | 0.031 | 0.439 |
| 20.0 | 0.5 | 0.031 | 0.440 |
|  | 1.0 | 0.016 | 0.220 |

*Source*: From Svergun et al. [2].

maximum $q$ values can be calculated geometrically from

$$q_{min(max)} = \frac{4\pi}{\lambda} \sin\left[\tan^{-1}\left(\frac{r_{min(max)}}{L}\right)\right]$$

where $r_{min}$ and $r_{max}$ are the distances from the detector centre to the nearest and furthest exposed pixels and $L$ is the sample-to-detector distance. This range can be extended by offsetting the detector. Note that from the definition of $q$, doubling $L$ (halving $2\theta$) has nearly the same effect as doubling the wavelength. Whether $L$ or $\lambda$ is changed depends on the collimation-dependent flux and the transmission at a given wavelength [2]. Table 3.4 lists typical $q_{min}$ and $q_{max}$ values applicable for SANS beamlines. Note that similar calculations may be performed for SAXS beamlines, although it is less common to overlap SAXS data at different sample-to-detector distances since on most SAXS beamlines the detector is held at fixed position, not on a rail in a large adjustable tank as on most SANS beamlines.

An excellent detailed account of SANS beamline instrumentation is available [20].

## 3.7  ULTRA-SMALL-ANGLE SCATTERING INSTRUMENTS

Both USAXS and USANS instruments are often based on the Bonse-Hart camera (Figure 3.11) USANS beamlines are available at several international neutron scattering facilities (Table 3.2), and a similar limited number of

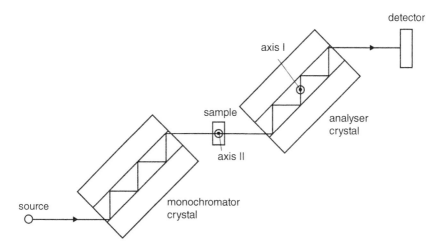

**Figure 3.11** Schematic of a Bonse-Hart camera. The standard setup involves rotation of the analyser crystal about rotation axis I, although alternative systems in which the detector and analyser are rotated together about rotation axis II in the sample have also been developed. *Source*: Redrawn from Pedersen [10]. © 2002, Elsevier.

USAXS instruments are installed at a few synchrotrons internationally (Table 3.1).

In the Bonse-Hart camera, the sample is placed between monochromator and analyser crystal, the latter being rotated (rotation axis I in Figure 3.11) to measure the rocking curve of x-ray or neutron intensity [10]. The intensity profile obtained is slit smeared. Since the beam is monochromated using channel-cut crystals, high angular resolution is achieved. Alternative Bonse-Hart systems involve rotation of the analyser crystal on a detector arm around an axis in the sample (rotation axis II in Figure 3.11). The monochromator and analyser are often triple-bounce Si(111) channel-cut crystals [21].

With a Bonse-Hart camera, it is possible to access scattering down to $q = 2 \times 10^{-5}$ Å$^{-1}$ [21], i.e. to probe systems with ordering on the lengthscale of 100 nm or more, such as porous and particulate materials or phase-separated polymer blends with inhomogeneities of this size (examples are discussed in Section 5.10 and Section 5.11). The Bonse-Hart system is used for isotropic samples.

USAXS can be performed using a Bonse-Hart camera in the same way as for USANS. For USAXS, the Bonse-Hart camera typically uses Si(111) or Si(220) channel-cut crystals crystals [22]. Usually USAXS is performed using 1D collimation, although 2D collimated USAXS has been developed [22]. The former leads to slit-smeared intensity profiles. USAXS is used to investigate systems containing large particles or porous materials or systems such as crystals with ~100 nm scale defects.

# 3.8 STANDARD SAMPLE ENVIRONMENTS – SAXS

A common type of holder for solution samples is an individual glass capillary for which there are a variety of holders available on different beamlines, with temperature control in many cases. X-ray capillaries are usually made of borosilicate glass (which has a low coefficient of thermal expansion). Typical capillary diameters are in the range 0.5–2 mm; however, microcapillaries are also available. Plastic microtubes designed to have a low scattering background for x-ray studies are also available, made from polycarbonate or another polymer. It has been pointed out that lower diameter capillaries (e.g. 0.9 mm compared to 1.7 mm) can reduce radiation damage in synchrotron BioSAXS measurements, as well as reducing sample volume requirements [23].

For powder and polymer film/melt samples, DSC (differential scanning calorimetry) pans can be convenient sample holders for temperature-dependent measurements using a relatively low sample volume. On hard x-ray SAXS beam lines (with $\lambda = 1$ Å, for example), aluminium DSC pans can be used without modification, as the transmission of x-rays is sufficiently high; however, on beamlines with longer wavelength, holes may be punched into the DSC pans and windows made of kapton or mica inserted either side of the sample in a sandwich arrangement (Figure 3.12). These DSC pans can be placed in a system such as a Linkam™ DSC instrument. These can provide simultaneous DSC measurements in single pan mode, provided the instrument is previously calibrated with a suitable standard. However, the system is often just used as a hot-stage with a wide temperature range, including sub-ambient when a liquid nitrogen cooler unit is attached. SANS/DSC has also been developed using a custom-designed aluminium crucible connected to a commercial DSC controller [24], although it is not yet widely exploited.

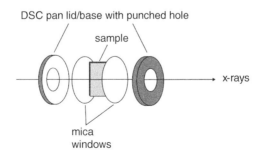

**Figure 3.12**  Expanded diagram of a modified DSC pan, which sandwiches a sample between a pair of mica windows within the base and lid of a DSC pan with punched holes.

A wide variety of designs of homemade cells with low background windows using materials such as mica, polyimide (Kapton™), polycarbonate, or aluminium (at low wavelength) have also been used on SAXS beamlines to contain more viscous or gel or soft solid samples sandwiched between the windows. It should be noted that mica is a polycrystalline material and can produce sharp Bragg peaks observed in WAXS patterns, but it has low background for SAXS [7]. Polyimide can give a broad peak corresponding to a spacing centred on $d = 25\,\text{Å}$, but this is usually weak and can be removed by background subtraction.

Position controllable multisample racks designed to fit standard sample cells such as capillaries or DSC pans are available on many synchrotron beamlines and some laboratory instruments. These are sometimes contained in temperature-controlled housings.

On BioSAXS synchrotron beamlines, the solution is usually delivered to a flow-through (open ended) capillary fixed in position in the beamline. This reduces variability in background scattering due to differences in wall thickness from capillary to capillary. The capillary has to be thoroughly washed, rinsed, and dried before loading a fresh sample. This is usually done under automated software control, by drawing washing solution, rinsing solution (usually water or a surfactant solution), and air into the capillary. The flow-through capillary may be filled robotically from a syringe delivering samples from an autosampler containing a multi-well plate as in the EMBL (European Molecular Biology Laboratory) BioSAXS robot used on several synchrotron BioSAXS beamlines. Figure 3.13 shows this system installed on a beamline at Diamond Light Source, UK. Typical sample volumes required for a series of 10–20 frames are 10–50 µl, depending on sample viscosity.

Radiation damage is a potential problem due to the high brilliance of the source at synchrotron SAXS beamlines. This leads to high absorption and ionization effects, which can cause aggregation. Ionization events in aqueous solution lead to the formation of hydroxyl ($OH^-$) or hydroperoxyl ($O_2H^-$) ions [2] also known as hydrated electrons. Flowing the sample during x-ray exposure is the best method to reduce beam damage effects, and this is becoming standard on bioSAXS beamlines where the sample is delivered automatically into a capillary and a fresh volume of sample is flowed into the beam for each frame of SAXS data recorded. A sequence of frames is measured with the shortest acquisition time that gives a reasonable not-too-noisy intensity profile (this can be ~100 ms–30 s depending on the beamline and the nature of the sample). The time-dependence of the signal recorded in each frame should then be monitored for evidence of changes due to x-ray exposure, which can indicate beam damage (in a sample that is not expected to evolve kinetically in this period). This is discussed further in Section 2.12. In fact, the software on many BioSAXS beamlines automatically

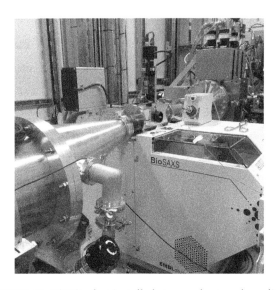

**Figure 3.13**  EMBL BioSAXS robot installed at a synchrotron beamline (B21 at Diamond Light Source, UK). *Source*: From Ref. [25].

checks for differences in intensity profiles from frame-to-frame and warns the user of potential radiation damage. Choice of buffer can also influence beam damage, since high salt content increases absorption and hence radiation damage. Addition of radical scavengers can be used to reduce radiation damage, although, of course, such reagents can influence the structure or assembly of the molecule/protein under investigation. The same consideration applies for added cryoprotectant molecules such as sugars or ethylene glycol [2]. In general, use of such potentially structure-perturbing additives should be avoided. Although in some synchrotron experiments such as single crystal XRD cryo-cooling techniques can be used to reduce radiation damage, for obvious reasons this cannot be employed for solution SAXS measurements or measurements on many soft materials. X-ray footprinting is a method using controlled radiolysis (i.e. 'beam damage') to generate $OH^-$ radicals to probe solvent accessibility of hence conformation and dynamics of macromolecules such as proteins and DNA in aqueous solution [26, 27]. The analysis is done by offline mass spectroscopic or sequencing methods. It has yet to be combined with SAXS although there seems to be potential for this if the radiolysis conditions can be controlled.

An alternative delivery system for BioSAXS samples uses a high-performance liquid chromatography (HPLC) system that enables biomolecules such as proteins in solution under flow to be analysed or separated by HPLC prior to delivery into the x-ray beam. The HPLC system may have MALS (multi-angle light scattering) or refractive index detection to further characterize the molecular weight of the species in the sample.

Size exclusion chromatography (SEC) is a type of HPLC used to separate macromolecules within a porous bead column, and it has been employed on synchrotron SAXS beamlines. SEC has also been used at SANS beamlines [28]. SEC-SAXS requires a typical sample volume ~ 50–100 µl [29]. Sample component separation using SEC is usually monitored with UV detectors, although static light scattering (MALS) is another option that is also available in commercial instruments.

## 3.9   STANDARD SAMPLE ENVIRONMENTS – SANS

The standard sample holder for liquid samples at SANS beamlines is a high-precision wall thickness quartz cuvette with a typical path length of 0.5–2 mm. The sample volume is typically 100–200 µl. Several designs of these are available (Banjo-shaped or rectangular, different types of stopper, etc.). Usually they are mounted in multi-sample racks under computer positional control for placement in the beam. The sample rack may have temperature control if required. Other types of samples (powders, polymer films, and melts, etc.) can be mounted in aluminium cans or other aluminium holders (e.g. wrapped in aluminium foil). Aluminium (Al) has a very low incoherent scattering cross section (Table 5.1) and also low absorption (Table 2.1). It is necessary to be careful after the measurement since metals, including Al, can be activated by the neutron beam and Al sample holders should be checked for radioactivity (this is usually done formally at a testing laboratory at the facility).

## 3.10   STANDARD SAMPLE ENVIRONMENTS – GISAS

Samples for GISAXS and GISANS are commonly analysed as films on solid substrates. Silicon wafers are a common choice of substrate, as they are available in very flat polished form. Silicon (actually silicon oxide at the surface) has a high electron density background for x-ray measurements but low scattering length density for neutrons (Table 6.1), so neutron reflectivity may be performed by passing the neutron beam through a Si substrate. Gold substrates are another common solid surface that can be prepared with a high degree of flatness. GISAS measurements may also be performed on films at the solution-air interface with the solution in a Langmuir trough, for example.

In synchrotron x-ray and neutron beamlines, GISAS is performed with the sample lying horizontal (this is obviously essential for measurements at the liquid interface!). A 2D detector collects the GISAS pattern. Reflectivity measurements are performed on dedicated beamlines, with the sample placed on a goniometer stage for alignment and particular types of intensity scans (discussed in Section 6.3) and a movable detector. Laboratory x-ray reflectivity instruments based on modified diffractometer designs can have the sample mounted vertically. Due to the high flux of synchrotron x-ray beams, beam damage can be particularly problematic for thin film samples of soft materials (e.g. burning stripes on the sample!), and this should be carefully monitored and avoided as necessary by beam attenuation.

Some specialized GISAS sample cells such as rheometers have been used in a limited number of studies. Controlled humidity cells are more common. Examples of studies with such sample environments are discussed in Section 6.6. This book does not cover surface science studies with GISAS/reflectivity in ultra-high vacuum environments.

# 3.11 MICROFOCUS SAXS AND WAXS

Synchrotron beamlines offer the possibility to focus the beam down to micron scale dimensions at the sample (down to $1\,\mu m \times 1\,\mu m$ [30]). This enables the possibility for scanning microfocus or microbeam analysis of the structure, ordering or texture across the sample. Figure 3.14 shows different types of optics that may be employed for microfocussing [31].

Microfocus SAXS (and SAXS/WAXS) has, for example, been applied to study the order in polymers, including polymer crystals, probing the order across a spherulite [31, 32] or to probe structure variations in regions of deformation such as around a crack tip [33, 34], scanning across the sample. Figure 3.15 shows an example of scanning WAXS patterns for a polymer spherulite (polyhydroxybutyrate) assembled into a map that shows changes in the crystal orientation (via orientation in the WAXS pattern), scanning across the sample with a $10\,\mu m \times 10\,\mu m$ beam [35, 36]. Similar compilation maps of SAXS patterns scanning across samples provide information on texture (nanoscale ordering) across the sample.

Microfocus (also known as microbeam) SAXS (and SAXS/WAXS) has been applied to many other systems as well as reviewed elsewhere [30, 36–38].

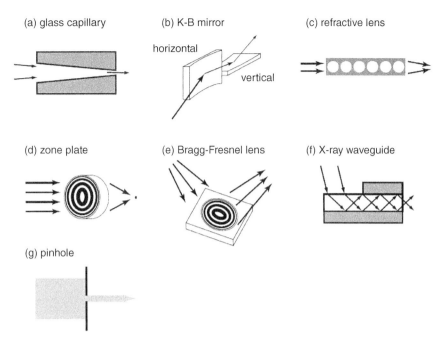

(a) glass capillary

(b) K-B mirror

horizontal

vertical

(c) refractive lens

(d) zone plate

(e) Bragg-Fresnel lens

(f) X-ray waveguide

(g) pinhole

Figure 3.14   Different types of optics for microfocus SAXS. *Source*: From Nozue et al. [31]. © 2007, Springer Nature.

## 3.12   SPECIALIZED SAMPLE ENVIRONMENTS

### 3.12.1   Stopped Flow

Stopped flow kinetics experiments are performed on many SAXS and SANS beamlines. This technique enables time-resolved measurements probing processes such as self-assembly in solution, with time resolution down to a few ms for SAXS and 100 ms for SANS, which requires longer mixing times (and counting times) [39–42]. The technique has been used to investigate the self-assembly of surfactants, colloids, and the formation of inorganic structures templated by organic species [40]. Section 5.9 discusses some examples of time-resolved SANS studies after stopped flow from amphiphilic systems. Stopped flow mixing is also valuable in SAS studies of protein folding and conformational changes in response to binding of ligands as well as changes in virus structures in response to solution conditions. Stopped-flow devices from BioLogic are used on several beamlines, since they have a short

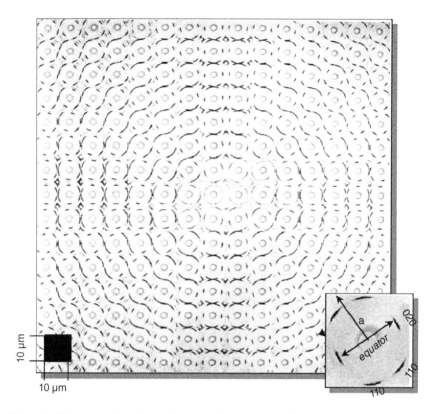

**Figure 3.15** Example of microfocus WAXS. Positional map of WAXS patterns located at corresponding positions scanning across a polymer spherulite. Inset: indexation of the peaks in the WAXS pattern. *Source*: Figure 2 from Mahendrasingam et al. [35].

dead-time and can be used for processes with kinetics in the range from ~1 ms up to several minutes. Rapid mixing in stopped flow cells can occur in 0.1 ms for samples with water-like viscosity; however, to avoid turbulence, the sample should be probed a few mm from the mixing point, increasing the dead time to a few ms [43]. Figure 3.16 shows such a device in situ on a SAXS beamline at the European Synchrotron Radiation Facility (ESRF) [43]. The dead time can be further reduced using continuous flow rather than stopped flow [43]. The mixing time can be studied by transmission measurements, e.g. using mixtures of water and KBr and measuring at the Br edge (Section 4.5) [43].

SAXS with stopped flow has been used to examine the kinetics of assembly of a variety of micellar and colloidal systems [39, 43, 44]. So-called stroboscopic SAXS was exploited with stopped flow kinetics in a study of the transition from spherical to wormlike micelles in an anionic surfactant

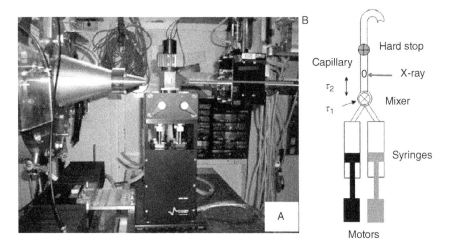

**Figure 3.16** (a) A Stopped flow cell in a SAXS beamline (ID02 at ESRF). (b) Schematic of the stopped flow device. The dead times $\tau_1$ and $\tau_2$ correspond, respectively, to the mixing time and transfer time to the observation capillary. Source: From Narayanan et al. [43]

solution [45]. The stroboscopic technique was required due to the slow read-out time of the CCD detector used (190 seconds). By strobing the data acquisition, a time resolution of 2.5 ms was achieved in the measurements (strobing involves repeat measurements of data acquired for different delay times after mixing before collection of the data frame). A geometric progression was used for each data frame. The same approach was used to study the unimer-micelle transition in a sugar-based surfactant system [46].

## 3.12.2   Rheology

Rheo-SAS refers to the combination of rheology (the study of the flow behaviours of matter) and SAS. Shear flow is applied with simultaneous measurement of rheological properties (typically shear modulus), although in some cases shear is applied to align the sample to provide additional structural information, without simultaneous rheology measurements. A considerable number of studies have been performed, involving the use of many different rheometer geometries. These are briefly summarized in the following. Rheo- and flow-SANS have been reviewed [47], as has rheo-SAXS [48]. Specialist reviews also cover topics including rheo-SAS studies of block copolymers, this includes in situ studies as well as many ex situ measurements on flow aligned samples [49].

## Couette Cells

This is the most common type of rheometer geometry at synchrotron beamlines, and commercial instruments adapted for SAXS and SANS are available at many facilities, following earlier pioneering developments [50–55], and experiments on many types of systems have been performed, as reviewed elsewhere [47, 56, 57]. The Couette geometry comprises concentric cylinders, with an inner stator and outer rotor. The gap is typically 0.5–2 mm, which is a good path length for SAS studies. This geometry is suited to the investigation of lower-viscosity fluids and has been used to investigate the shear-induced alignment of colloid and surfactant assemblies, polymer solution structures, liquid crystals, and biomolecular assemblies. The sample is generally subjected to continuous shear, providing a shear rate-dependent measure of the viscosity, denoted $\eta(\dot{\gamma})$, where $\eta$ is viscosity and $\dot{\gamma}$ is the shear rate. Some rheometers are also capable of oscillatory measurements in the Couette geometry, providing the frequency and shear (or strain) amplitude-dependent dynamic shear moduli $G'$ (elastic or storage modulus) and $G''$ (loss modulus). Two configurations are possible, providing access to different planes with respect to the shear flow, as shown in Figure 3.17. The radial configuration provides SAS patterns in the $(\mathbf{v}, \mathbf{e})$ plane whereas passing the beam tangentially, the alignment in the $(\nabla \mathbf{v}, \mathbf{e})$

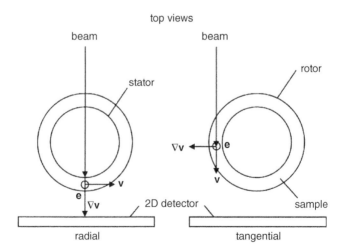

**Figure 3.17** Two configurations of Couette cells with the beam incident radially or tangentially provide access to different planes with respect to the shear flow. Here $\mathbf{v}$, $\nabla \mathbf{v}$, and $\mathbf{e}$ denote the shear flow, shear gradient and neutral directions, respectively. *Source*: From Hamley et al. [58]. © 2017, American Chemical Society.

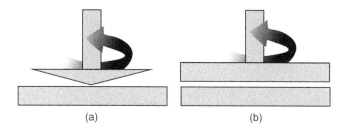

(a)                                        (b)

Figure 3.18   Schematic of (a) cone-and-plate and (b) plate–plate rheometers.

plane is probed. By rotating the Couette cell so that the rotation axis lies horizontally, it is possible to access the other shear plane (the $\mathbf{v}$, $\nabla\mathbf{v}$) plane [47, 55, 59, 60]. Couette cells have been constructed from polycarbonate for SAXS (since it has relatively high transmission and low background) and aluminium may be used for a rheo-SANS geometry.

## Plate Rheometers

Cone-and-plate cells may be used to investigate gels and soft solids, while plate–plate geometries are suited to stiffer solids and can be used for polymer melts, etc. These geometries are shown in Figure 3.18. Using another type of rheometer with a vertical geometry and oscillatory deformation, a commercial instrument with a plate–plate geometry was adapted for in situ rheo-SAXS and rheo-SANS. The commercial rheometer plates were modified with Kapton-covered windows for SAXS (Figure 3.19) [61] or replaced with polycarbonate for SAXS or aluminium plates for SANS [62, 63]. The instrument was used to study a number of block copolymer systems (see e.g. [61, 62, 64]). In an alternative configuration, the x-ray beam is displaced vertically (using a single crystal mirror) through horizontal polyimide parallel plates (with thinner window areas) in a commercial rheometer [65]. This was used to study block copolymer alignment via in situ synchrotron SAXS. Linkam manufactures a plate–plate shear cell that comprises a rotating bottom plate and fixed upper plate. This device has been used by several groups, e.g. in rheo-SAXS/WAXS studies of shear-induced alignment of crystal 'shish-kebab' precursor structures in polymers; for other selected examples, see refs. [66–68]. The standard optical quartz windows are replaced with Kapton windows for SAXS/WAXS studies [66]. Similar home-built systems have been used for SAXS and SANS studies of materials under shear [48, 69, 70]. Several groups developed sliding plate (reciprocating) shear cells for in situ or ex situ SAXS or SANS measurements [71–80].

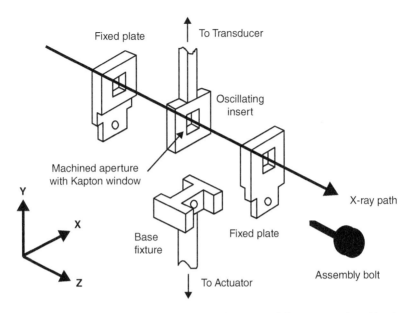

**Figure 3.19**   Parallel plate oscillatory shear device used for in situ SAXS/rheology. The plates fit in a commercial soft solids rheometer. *Source*: From Pople et al. [61]. © 1990, American Chemical Society.

## Continuous Flow

Continuous flow through a capillary can provide alignment to susceptible samples such as concentrated surfactant, polymer or colloid solutions as well as biomolecular structures such as peptide and protein amyloid or lipopeptide structures. Since samples are delivered into capillaries by continuous flow at BioSAXS beamlines, 2D SAXS patterns can be checked for samples likely to show alignment. Many groups also developed capillary and parallel plate (Poiseuille) flow systems or extensional flow devices for more systematic orientation studies by SAXS [47, 51, 81–83] or SANS [47, 83, 84]. Continuous flow is also used in microfluidic SAS, discussed further in Section 3.12.6.

## Other Flow Systems

Various compression, roll-casting and extrusion flow methods have been used to produce aligned samples of polymers for SAXS (or SAXS/WAXS) studies, especially to examine block copolymer alignment, but these are not generally employed for in situ studies. Details are provided elsewhere [85, 86]. An opposing jet (crossed slot) geometry has also been used to generate extensional flow for SAXS [87] and SANS [88] experiments [47].

## 3.12.3   Tensile Deformation

This can be considered to be an extension (pun intended!) of rheo-SAS in the case that the stress–strain behaviour of the material is measured as well as delivering in situ tensile deformation. The mechanical properties may not be recorded and the device may simply be used to controllably stretch a material. A number of commercial tensile testing devices have been installed at synchrotron beamlines, including the (now-discontinued) miniature uniaxial tensile tester from Linkam, as well as larger-scale commercial biaxial tensile instruments.

SANS with in situ uniaxial extensional deformation has been applied to study alignment in plasticized PVC films and block copolymer elastomers [89]. SAXS with in situ tensile deformation has been used to examine alignment and structural transitions in many types of elastomeric polymer and copolymer [90]. In another example, time-resolved SAXS on copolymer films subjected cyclical stretching has also been used to examine materials designed as potential heart valve replacements [91].

Stretching has also been employed in SAXS studies of muscle structure, however contraction is generally stimulated electrically as discussed further in Section 3.12.7.

## 3.12.4   Pressure Cells

Materials, including inorganic and soft materials, often exhibit novel phase behaviour under conditions of high pressure, and proteins and other biomolecules show interesting folding behaviour under these conditions. Many groups have thus developed pressure cells to investigate these effects with in situ small-angle scattering [92–106]. A number of designs of pressure cells are available. In order to support the pressure within the cell, a strong window material is required. This is usually beryllium or diamond for SAXS or sapphire for SANS measurements [101]. One common design is the diamond anvil cell, in which a sample is compressed between the tips of a pair of diamond crystals. These can deliver pressures up to hundreds of GPa. However, they can cause problems due to path length changes under pressure and so are not usually used for measurements on biomacromolecular and soft matter samples [102].

Diamond anvil cells are also not suitable for pressure jump experiments. A number of designs of cells for pressure jump experiments are available, achieving hydrostatic pressure changes of hundreds of MPa (1000 bar+) in 1–5 ms [92, 98, 99, 104, 105]. Smaller pressure jumps of 200 bar can be achieved in sub-ms timescales, and this has been used, for example, in synchrotron SAXS measurements of the effect of pressure on nucleotide binding to a myosin protein fragment [107]. Pressure-jump SANS experiments include, e.g. a study of pressure-induced gelation of $\beta$-lactoglobulin subjected to a pressure jump from 0.1 to 315 MPa in a few minutes [108].

## 3.12.5    Temperature Jump

SAXS temperature jump (T-jump) experiments have been performed for more than 30 years [109–113], and this type of measurement has been and is being done at many synchrotron SAXS beamlines. Usually, fast T-jump experiments are too rapid for time-resolved SANS studies. The most common T-jump apparatus involves the use of a high-power pulsed infrared or visible laser illuminating small sample volumes. Small T-jumps for solution samples can be done in µs, being limited by thermal diffusion. However, larger temperature jumps lead to hydrodynamic flows, and so are difficult to perform in a controlled fashion [43]. Laser flash photolysis can be carried out for samples containing photoactive groups. Microwave heating at 2.45 GHz has also been used for aqueous phospholipid solution samples, with heating rates up to 1.4 K s$^{-1}$ (and a higher cooling rate) reported [112]. Heat guns (with temperature-control feedback) have also been employed, producing usable heating rates for a sample in a 1 mm diameter capillary of 50 °C in 15 seconds, which was exploited to study phospholipid phase transitions [109]. Ohmic heating can produce T-jumps of 10 K (ms)$^{-1}$, but is complicated by potential electrical breakdown and gas generation at the electrodes [110].

Fast heating and cooling rates are available with Linkam DSC and capillary hot stages with controlled liquid nitrogen cooling. The instrument specifications indicate rates of up to 150 °C/min for heating, and 100 °C/min for cooling (using the liquid nitrogen cooling pump). The BioLogic stopped flow system mentioned in Section 3.12.1 has an optional temperature-jump attachment, which can achieve 40 K T-jumps in several ms.

Systems to achieve temperature jumps by moving the sample between hot and cold stages have been developed, such as the rapid heat and quench cell translation system on the Quokka SANS beamline in Australia [42]. Rapid cooling rates up to $1.5 \times 10^4$ K s$^{-1}$ have been achieved in studies of electric-arc generated alloy particles quenched onto cool copper blocks in XRD studies [114], and this has potential in SAXS investigations as well.

### 3.12.6 Microfluidic Devices

Microfluidic techniques enable SAXS experiments to be performed on low volumes (1 nl or less) of sample and are thus proving extremely useful in studies such as high throughput analysis of proteins in solution [115]. There are many designs of microfluidics cells [116, 117] in two main classes – multichannel and single-channel configurations (Figure 3.20). Multichannel devices are used for example for phase mapping studies since composition gradients can be created with low sample volumes, enabling the mapping of phase diagrams via the generation of concentration gradients using hydrodynamic laminar flow focussing (among other mixing flows) and scanning along the microfluidic channel [116].

Single-channel microfluidics can be used to examine a sample with nanolitre volume, including flow effects. As shown in Figure 3.20b, the region where the beam passes may be designed to have a shorter path length and in this region, reduced transmission windows may be inserted [116]. The lower flux at SANS beamlines means that single channel microfluidics with good time resolution is only possible for concentrated, ordered samples [117].

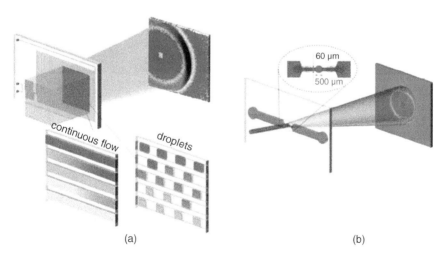

(a)                                      (b)

**Figure 3.20** (a) Multichannel microfluidics, (b) Single-channel microfluidics. In part (b), 500 μm corresponds to the typical size of a SANS beam in scanning flow-SAS studies and 60 μm is a typical microfluidic channel diameter. *Source*: From Lopez et al. [117]. © 2018, International Union of Crystallography.

Droplet microfluidics where samples are dispersed in droplets within a continuous flow of immiscible liquids, can further be used to reduce sample volume requirements [118]. This method has been used in SAXS studies, such as calcite crystallization (examined by SAXS/WAXS) [119] and protein crystallization using protein solution droplets suspended in different oils [120].

SAXS microfluidic cells are typically made from polymeric materials such as polystyrene, Kapton, poly(dimethylsiloxane) (PDMS), or cyclic olefin (co-)polymers, depending on the requirements of solvent and temperature tolerance (and evaporation), biocompatibility and avoidance of surface absorption, background/transmission and beam damage considerations [116, 117]. Other designs such as silicon-on-glass microreactors have been developed [119]. Microfluidic SAXS studies have been reviewed, including both high throughput parameter mapping studies (protein concentration series, lipid and polymer phase mapping, etc.) and single channel mixing kinetic studies [116]. Some systems incorporate additional detection capabilities such as UV absorbance measurements.

Microfluidics has also been used with SANS to study the flow behaviour mechanisms and kinetics of polymers and other molecules under flow [117, 121]. The choice of cell window materials has been carefully analysed, considering inorganic glasses, polymers or (for SANS specifically) metals and also evaluating the requirements for solvent/temperature stability, biocompatibility as well as transmission and background scattering [117]. Polymers such as PDMS are commonly used as inexpensive materials for microfluidic cell fabrication, enabling rapid prototyping and good biocompatibility, although it has high absorption and background for neutrons. Kapton is a good choice for both SAXS and SANS [117].

## 3.12.7 Electrical and Magnetic Fields

Several designs of electrical cells have been developed for use with in situ SAS experiments including both AC and DC cells, and configurations with the electric field applied perpendicular (most common, Figure 3.21) or parallel to the direction of the beam. These devices have been used in synchrotron SAXS studies of the alignment of dielectric materials including block copolymer melts, inorganic nanoparticles, elastomers, and biomolecular structures [48].

Muscle contraction can be stimulated electrically, and this has been examined by in situ SAXS using custom designed cells with platinum electrodes placed either side of the muscle. Single muscle fibres are preferred to whole muscle, although they are more difficult to prepare and handle.

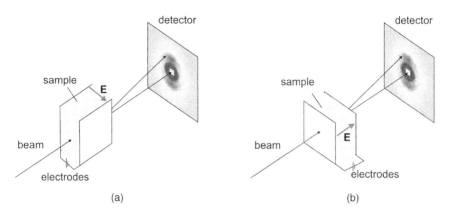

**Figure 3.21**  Configurations of cells that apply electric fields (a) perpendicular, (b) parallel to a beam.

High-frequency pulses (e.g. at 50 Hz) produce sustained contraction (termed tetanic contraction) [122]. The recovery period is 2–3 minutes [122].

Muscle has a striated structure arising from a hierarchical organization, some of which is on a length scale that can be probed by SAXS, specifically the sarcomere structure within myofibrils, which arises from overlapping actin and myosin filaments (Figure 3.22).

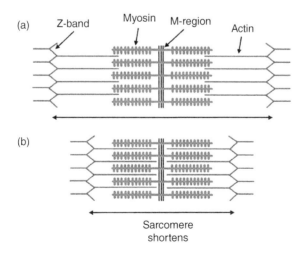

**Figure 3.22**  Structure within the sarcomere structure of muscle filaments at rest (top) and under contraction (bottom). *Source*: Based on Harford et al. [122]. © 1997, IOP Publishing Ltd.

The myosin headgroups are arranged helically, with a period for the three coaxial helices in myosin heads of 42.9 nm (M1) in vertebrate striated (skeletal and cardiac) muscle [122, 123]. The SAXS pattern comprises a series of layer lines (Figure 3.23) of which the third order, M3, is particularly pronounced (Figure 3.23). The M1 period is slightly increased upon stimulation in the active state, but more significant changes are observed in that off-meridional reflections weaken or disappear (Figure 3.23) [123].

Actin is also a helically ordered filament, with a pitch of about 37.5 nm (i.e. also in the range of length scales that can be probed by SAXS) [123]. Tropymyosin molecules lie in the groove between pairs of actin filaments, and at the end each tropomyosin binds to troponin, which has a repeat spacing along the filament in the resting state of 38.5 nm [123]. The layer line reflections in Figure 3.23 have fine structure, which arises from cross-interference effects between mirror structures within the sarcomere (Figure 3.22), with contributions from C-protein and titin (proteins that can bind to actin or myosin filaments) array periodicities, i.e. the reflections are sampled by a longer superstructure from arrays of myosin head triplets, with a periodicity of 704 nm [122, 123]. Muscle structure has even been studied by in situ SAXS on a tethered flying insect (fruit fly) [124].

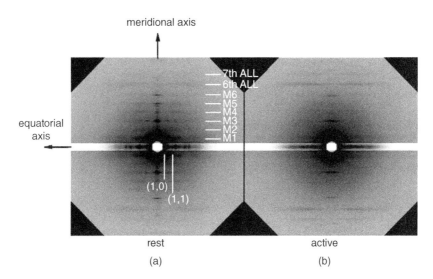

**Figure 3.23** SAXS patterns from frog skeletal muscle (a) at rest, (b) during isometric contraction. *Source*: From Reconditi [123].

Electromagnets and high field permanent magnets have been used for in situ SAXS and SANS studies at many synchrotron and neutron beam lines as a tool to align diamagnetic materials such as block copolymers and liquid crystal polymers [48]. Ex situ measurements on prealigned samples have been performed, as well as in situ studies. Examples include in situ SAXS studies of block copolymer alignment during cooling (from the isotropic phase) in high magnetic fields [125–128], and in situ WAXS (with ex situ SAXS) of the ordering of a block copolymer containing a smectic-forming liquid crystal block [129]. Examples of SANS studies include an investigation of the conformation of an side-chain liquid crystal polymer blend aligned in an electromagnetic field [130], and in situ magnetic field-induced alignment of a side group liquid crystal block copolymer [131].

## 3.12.8 Controlled Humidity

Environmental chambers that provide control over relative humidity have been developed for in situ SAXS studies on ionomer membranes and polymerized ionic liquids [132–136]. Simultaneous conductivity and SAXS measurements have been performed under conditions of controlled humidity, to investigate the properties of polymer electrolyte membranes [137]. Neutron fibre diffraction has been performed on DNA under controlled hydration conditions [138].

## 3.12.9 Gas-Phase SAS

The formation of particles in diffusion flames (where fuel and oxidiser mix diffusively) has been studied with in situ SAXS [121]. The height above the burner is a measure of residence time and can be used as a proxy in kinetics studies. Use of a synchrotron in the SAXS experiment is essential due to the low concentration of particles in the vapour phase. In one example, the nucleation and growth of soot particles in an acetylene flame has been studied using SAXS combined with USAXS [139], and the formation of silica particles from hexamethyldisiloxane vapour in a co-annular stream with combustion gases has similarly been examined [140]. In these studies, the size and fractal dimensions of the particles were analysed.

Flames produced by spraying from a nozzle produce higher degrees of supersaturation and in this case height above the nozzle is used as a measure of residence time. SAXS studies using such jet flames include an investigation of zirconia particle formation [121]. The formation of aerosol nanodroplets of water/$D_2O$ have been investigated by in situ SAXS [141] and SANS [142] using supersonic nozzles, enabling the determination of nucleation rates and particle sizes.

SAXS has been performed on inorganic oxide particles in an aerodynamic levitation furnace (which relies on high power $CO_2$ laser heating of sample in an argon gas stream [143]) in an in situ study of phase transitions under contactless conditions [144]. Acoustic (ultrasound) systems have been developed to produce levitated aerosol droplets (0.06–2 mm diameter) for SAXS and WAXS studies using small beams [145, 146]. This enables the study of these droplets as free particles, in a contactless environment [146]. Levitation cells that also enable in situ Raman spectroscopy have been developed [146, 147]; Figure 3.24 shows an example.

SAXS on acoustically levitated droplets has been used to study processes associated with evaporation from the droplets, including the agglomeration of model proteins [145], to model spray drying of inorganic oxides, [147] and to examine nanoparticle assembly into mesocrystals [148]. In another example, acoustic levitation SAXS was used to study droplets of oleic acid with internal liquid crystal (hexagonal and cubic micellar structures). The effects of hydration and ozonolysis were also examined [146]. Gold nanoparticle formation in acoustically levitated droplets of precursor solutions has been studied by SAXS combined with XANES (X-ray absorption near-edge spectroscopy) [149].

In related work, SAXS has been used to probe the kinetics of silver or gold nanoparticle formation in a liquid subjected to high-power laser ablation, which leads to bubble cavitation [150, 151]. The formation of nanoparticles in the cavitation bubbles was monitored by time-resolved SAXS.

## 3.12.10   Simultaneous Spectroscopic Measurements

SAXS/WAXS measurements have been combined with simultaneous FTIR spectroscopy to investigate polyurethane formation and phase separation, using an ATR FTIR instrument adapted to allow transmission of the x-ray beam [152, 153]. FTIR in transmission mode using a modified mini-spectrometer has also been combined with SAXS/WAXS on a polymer subjected to in situ extension in a study on a stress-induced polymer crystal phase transition [154]. SAXS/WAXS/Raman measurements have been demonstrated at several synchrotron beamlines [153, 155–159]. This technique is simpler to set up than FTIR, since Raman is a scattering

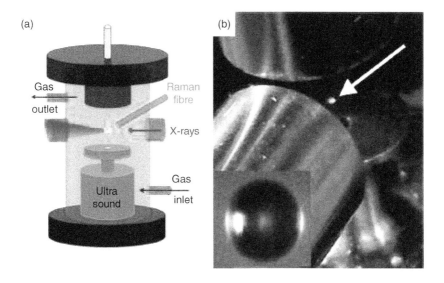

**Figure 3.24** Acoustic levitation of droplets for in situ SAXS (with Raman spectroscopy), (a) schematic, (b, c) photographs of levitated droplets. An 80 μm droplet (arrowed) (b) is shown in situ in a synchrotron beamline, and (c) with 532 nm laser illumination for Raman spectroscopy. *Source*: From Pfrang et al. [146]. Licensed under CC BY 4.0.

method (fibre optics are used to for the incident and scattered Raman laser light). As well as SAXS/Raman, a system that can additionally perform simultaneous UV-vis spectroscopic analysis has been demonstrated [159]. Figure 3.25 shows a schematic of the setup.

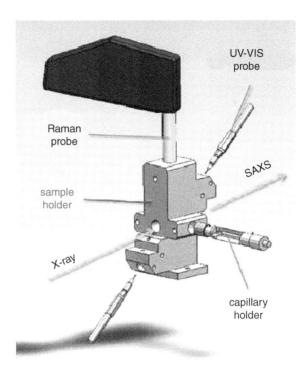

**Figure 3.25** Design for a simultaneous SAXS/Raman/UV-vis instrument at a synchrotron beamline. *Source*: From Haas et al. [159]. © 2014, American Chemical Society.

Experiments combining both UV-vis and x-ray fluorescence measurements with SAXS have been reported [160]. Dielectric spectroscopy has also been combined with SAXS/WAXS, for the investigation of polymer crystallization [161–163]. Dielectric spectroscopy provides information on the dynamics of the amorphous phase. SAXS/WAXS has also been combined with EXAFS (x-ray absorption fine structure) at a synchrotron beamline to investigate, for example, nanoparticle synthesis and the thermal synthesis of inorganic catalytic materials [164]. Simultaneous XANES/SAXS measurements have also been used to examine nanoparticle growth [149].

## 3.12.11   Simultaneous SAXS and Light Scattering

Synchrotron SAXS/WAXS has been combined with in situ light scattering to investigate polymer crystallization [156, 165, 166]. A light scattering system that can be adapted for SAXS has been used to examine polymerization mechanisms, including those occurring in a high-pressure supercritical $CO_2$ cell [167].

# REFERENCES

1. Wikipedia. (2020). List of synchrotron radiation facilities. https://en.wikipedia.org/wiki/List_of_synchrotron_radiation_facilities
2. Svergun, D.I., Koch, M.H.J., Timmins, P.A., and May, R.P. (2013). *Small Angle X-Ray and Neutron Scattering from Solutions of Biological Macromolecules*. Oxford: Oxford University Press.
3. Duke, P.J. (2000). *Synchrotron Radiation. Production and Properties*. Oxford: Oxford University Press.
4. American Light Source. (2020). https://als.lbl.gov/beamlines.
5. Cammarata, M., Eybert, L., Ewald, F. et al. (2009). Chopper system for time resolved experiments with synchrotron radiation. *Review of Scientific Instruments* 80: 10.
6. Diamond Light Source. (2020). www.diamond.ac.uk/Instruments/Soft-Condensed-Matter/small-angle/I22/specs.html.
7. Narayanan, T. (2014). Small-angle scattering. In: *Structure from Diffraction Methods* (eds. D.W. Bruce, D. O'Hare and R.I. Walton). Wiley Online Library: Wiley.
8. Roe, R.-J. (2000). *Methods of X-Ray and Neutron Scattering in Polymer Science*. New York: Oxford University Press.
9. Stribeck, N. (2007). *X-Ray Scattering of Soft Matter*. Berlin: Springer-Verlag.
10. Pedersen, J.S. (2002). Instrumentation for small-angle X-ray and neutron scattering. In: *Neutrons, X-Rays and Light Scattering Methods Applied to Soft Condensed Matter* (eds. P. Lindner and T. Zemb). Amsterdam: Elsevier.
11. Ungar, G. (1994). X-ray studies using synchrotron radiation. In: *Characterization of Solid Polymers: New Techniques and Developments* (ed. S.J. Spells). London: Chapman and Hall.
12. Institut Laue Langevin. (2020). https://www.ill.eu/users/instruments/instruments-list/d22/description/instrument-layout.
13. Higgins, J.S. and Benoît, H.C. (1994). *Polymers and Neutron Scattering*. Oxford: Oxford University Press.
14. Otomo, T. (2011). Small-Angle Neutron Scattering. In: *Neutrons in Soft Matter* (eds. T. Imae, T. Kanaya, M. Furusaka and N. Torikai). Wiley Online Library: Wiley.
15. Kawai, T. (1999). Multilayer neutron mirrors and their applications. *Acta Physica Polonica A* 96: 19–32.
16. Wilson, C.C. (2003). Time-of-flight neutron diffraction. In: *Neutron Data Booklet* (eds. A.-J. Dianoux and G. Lander). Grenoble, France: Institut Laue-Langevin.
17. Mortensen, K. (2011). Small-angle neutron scattering. In: *Neutrons in Soft Matter* (eds. T. Imae, T. Kanaya, M. Furusaka and N. Torikai). Wiley Online Library: Wiley.
18. Anderson, I.S. (2003). Neutron optics. In: *Neutron Data Booklet* (eds. A.-J. Dianoux and G. Lander). Grenoble, France: Institut Laue-Langevin.
19. Pedersen, J.S. (1995). Instrumentation for small-angle scattering. In: *Modern Aspects of Small-Angle Scattering*. NATO ASI Series C. 451 (ed. H. Brumberger), 57–91. Dordrecht: Kluwer.
20. Melnichenko, Y.B. (2016). *Small-Angle Scattering from Confined and Interfacial Fluids*. Berlin: Springer-Verlag.
21. Agamalian, M. (2011). Bonse-Hart USANS Instrument. In: *Neutrons in Soft Matter* (eds. T. Imae, T. Kanaya, M. Furusaka and N. Torikai). Wiley Online Library: Wiley.
22. Ilavsky, J., Jemian, P.R., Allen, A.J. et al. (2009). Ultra-small-angle X-ray scattering at the advanced photon source. *Journal of Applied Crystallography* 42: 469–479.
23. Schroer, M.A., Blanchet, C.E., Gruzinov, A.Y. et al. (2018). Smaller capillaries improve the small-angle X-ray scattering signal and sample consumption for biomacromolecular solutions. *Journal of Synchrotron Radiation* 25: 1113–1122.
24. Pullen, S.A., Booth, N., Olsen, S.R. et al. (2014). Design and implementation of a differential scanning calorimeter for the simultaneous measurement of small angle neutron

scattering. *Measurement Science and Technology* 25: 8.

25. Diamond Light Source. (2020). www.diamond.ac.uk/industry/Industry-News/Latest-News/Synchrotron-Industry-Focus-BioSAXS.html.

26. Xu, G.H. and Chance, M.R. (2007). Hydroxyl radical-mediated modification of proteins as probes for structural proteomics. *Chemical Reviews* 107: 3514–3543.

27. Gupta, S., Celestre, R., Petzold, C.J. et al. (2014). Development of a microsecond X-ray protein footprinting facility at the Advanced Light Source. *Journal of Synchrotron Radiation* 21: 690–699.

28. Jordan, A., Jacques, M., Merrick, C. et al. (2016). SEC-SANS: size exclusion chromatography combined in situ with small-angle neutron scattering. *Journal of Applied Crystallography* 49: 2015–2020.

29. Graewert, M.A. and Jeffries, C.M. (2017). Sample and buffer preparation for SAXS. In: *Biological Small Angle Scattering: Techniques, Strategies and Tips. Advances in Experimental Medicine and Biology* (ed. B. Chaudhuri), 11–30. Berlin: Springer.

30. Riekel, C., Burghammer, M., Davies, R. et al. (2009). Fundaments of soft condensed matter scattering and diffraction with microfocus techniques. In: *Applications of Synchrotron Light to Scattering and Diffraction in Materials and Life Sciences*. Lecture Notes in Physics. 776 (eds. T.A. Ezquerra, M.C. GarciaGutierrez and A. Nogales), 91–104. Berlin: Springer-Verlag Berlin.

31. Nozue, Y., Shinohara, Y., and Amemiya, Y. (2007). Application of microbeam small- and wide-angle X-ray scattering to polymeric material characterization. *Polymer Journal* 39: 1221–1237.

32. Kolb, R., Wutz, C., Stribeck, N. et al. (2001). Investigation of secondary crystallization of polymers by means of microbeam X-ray scattering. *Polymer* 42: 5257–5266.

33. Zafeiropoulos, N.E., Davies, R.J., Roth, S.V. et al. (2005). Microfocus X-ray scattering scanning microscopy for polymer applications. *Macromolecular Rapid Communications* 26: 1547–1551.

34. Lorenz-Haas, C., Muller-Buschbaum, P., Wunnicke, O. et al. (2003). Scanning microfocus small-angle X-ray scattering: a new tool to investigate defects at polymer-polymer interfaces. *Langmuir* 19: 3056–3061.

35. Mahendrasingam, A., Martin, C., Fuller, W. et al. (1995). Microfocus X-ray-diffraction of spherulites of Poly-3-Hydroxybutyrate. *Journal of Synchrotron Radiation* 2: 308–311.

36. Riekel, C. (2000). New avenues in x-ray microbeam experiments. *Reports on Progress in Physics* 63: 233–262.

37. Riekel, C., Burghammer, M., and Muller, M. (2000). Microbeam small-angle scattering experiments and their combination with microdiffraction. *Journal of Applied Crystallography* 33: 421–423.

38. Paris, O. (2008). From diffraction to imaging: new avenues in studying hierarchical biological tissues with x-ray microbeams (review). *Biointerphases* 3: FB16–FB26.

39. Panine, P., Finet, S., Weiss, T.M., and Narayanan, T. (2006). Probing fast kinetics in complex fluids by combined rapid mixing and small-angle X-ray scattering. *Advances in Colloid and Interface Science* 127: 9–18.

40. Grillo, I. (2009). Applications of stopped-flow in SAXS and SANS. *Current Opinion in Colloid & Interface Science* 14: 402–408.

41. Gradzielski, M. (2012). Dynamics of self-assembled systems studied by neutron scattering: current state and perspectives. *European Physical Journal: Special Topics* 213: 267–290.

42. Mata, J.P., Hamilton, W.A., and Gilbert, E.P. (2012). Application of time-resolved small angle neutron scattering to non-equilibrium kinetic studies. In: *Dynamics of Soft Matter Neutron Applications* (eds. V. García-Sakai, C. Alba-Simionesco and S.-H. Chen), 289–318. Berlin: Springer.

43. Narayanan, T., Gummel, J., and Gradzielski, M. (2014). Probing the self-assembly of unilamellar vesicles using time-resolved SAXS. In: *Advances in Planar Lipid Bilayers and*

*Liposomes*, vol. 20 (eds. A. Iglic and C.V. Kulkarni), 171–196. San Diego: Elsevier Academic Press Inc.

44. Narayanan, T., Wacklin, H., Konovalov, O., and Lund, R. (2017). Recent applications of synchrotron radiation and neutrons in the study of soft matter. *Crystallography Reviews* 23: 160–226.

45. Jensen, G.V., Lund, R., Gummel, J. et al. (2014). Monitoring the transition from spherical to polymer-like surfactant micelles using small-angle x-ray scattering. *Angewandte Chemie International Edition* 53: 11524–11528.

46. Jensen, G.V., Lund, R., Gummel, J. et al. (2013). Direct observation of the formation of surfactant micelles under nonisothermal conditions by synchrotron SAXS. *Journal of the American Chemical Society* 135: 7214–7222.

47. Eberle, A.P.R. and Porcar, L. (2012). Flow-SANS and Rheo-SANS applied to soft matter. *Current Opinion in Colloid and Interface Science* 17: 33–43.

48. Feng, J., Kriechbaum, M., and Liu, L. (2019). In situ capabilities of small angle x-ray scattering. *Nanotechnology Reviews* 8: 352–369.

49. Hamley, I.W. (2001). Amphiphilic diblock copolymer gels: the relationship between structure and rheology. *Philosophical Transactions of the Royal Society of London* 359: 1017–1044.

50. Lindner, P. and Oberthur, R.C. (1984). Apparatus for the investigation of liquid-systems in a shear gradient by small-angle neutron-scattering (SANS). *Revue de Physique Appliquee* 19: 759–763.

51. Hayter, J.B. and Penfold, J. (1984). Use of viscous shear alignment to study anisotropic micellar structure by small-angle neutron-scattering. *Journal of Physical Chemistry* 88: 4589–4593.

52. Ottewill, R.H. and Rennie, A.R. (1990). Scattering studies on polymer colloids under shear. *International Journal of Multiphase Flow* 16: 681–690.

53. Lindner, P. (1991). Small-angle neutron-scattering studies of liquid-systems in nonequilibrium. *Physica A: Statistical Mechanics and its Applications* 174: 74–93.

54. Penfold, J., Staples, E., and Cummins, P.G. (1991). Small-angle neutron-scattering investigation of rodlike micelles aligned by shear-flow. *Advances in Colloid and Interface Science* 34: 451–476.

55. Safinya, C.R., Sirota, E.B., and Plano, R.J. (1991). Nematic to smectic-a phase transition under shear flow: a non-equilibrium synchrotron x-ray study. *Physical Review Letters* 66: 1986–1989.

56. Rennie, A.R. and Clarke, S.M. (1996). Scattering by complex fluids under shear. *Current Opinion in Colloid and Interface Science* 1: 34–38.

57. Butler, P.D. (1999). Shear induced structures and transformations in complex fluids. *Current Opinion in Colloid and Interface Science* 4: 214–221.

58. Hamley, I.W., Burholt, S., Hutchinson, J. et al. (2017). Shear alignment of bola-amphiphilic arginine-coated peptide nanotubes. *Biomacromolecules* 18: 141–149.

59. Pople, J.A., Hamley, I.W., and Diakun, G.P. (1998). An integrated Couette system for *in situ* shearing of polymer and surfactant solutions and gels with simultaneous small-angle x-ray scattering. *Review of Scientific Instruments* 69: 3015–3021.

60. Calabrese, M.A. and Wagner, N.J. (2017). New insights from Rheo-small-angle neutron scattering. In: *Wormlike Micelles: Advances in Systems, Characterisation and Applications*, RSC Soft Matter Series, vol. 6 (eds. C.A. Dreiss and Y. Feng), 193–235. Cambridge: Royal Soc Chemistry.

61. Pople, J.A., Hamley, I.W., Fairclough, J.P.A. et al. (1997). Ordered phases in aqueous solutions of diblock oxyethylene/oxybutylene copolymers investigated by simultaneous small-angle X-ray scattering and rheology. *Macromolecules* 30: 5721–5728.

62. Hamley, I.W., Pople, J.A., Fairclough, J.P.A. et al. (1998). Effect of shear on cubic phases in gels of a diblock copolymer. *Journal of Chemical Physics* 108: 6929–6936.

63. Mortensen, K., Theunissen, E., Kleppinger, R. et al. (2002). Shear-induced morphologies of cubic ordered block copolymer micellar networks studied by in situ small-angle neutron scattering and rheology. *Macromolecules* 35: 7773–7781.

64. Hamley, I.W., Pople, J.A., Fairclough, J.P.A. et al. (1998). Shear-induced orientational transitions in the body-centered cubic phase of a diblock copolymer gel. *Macromolecules* 31: 3906–3911.

65. Meins, T., Hyun, K., Dingenouts, N. et al. (2012). New insight to the mechanism of the shear-induced macroscopic alignment of Diblock copolymer melts by a unique and newly developed Rheo-SAXS combination. *Macromolecules* 45: 455–472.

66. Somani, R.H., Hsiao, B.S., Nogales, A. et al. (2000). Structure development during shear flow-induced crystallization of i-PP: in-situ small-angle x-ray scattering study. *Macromolecules* 33: 9385–9394.

67. Somani, R.H., Yang, L., Hsiao, B.S. et al. (2002). Shear-induced precursor structures in isotactic polypropylene melt by in-situ rheo-SAXS and rheo-WAXD studies. *Macromolecules* 35: 9096–9104.

68. Yang, L., Somani, R.H., Sics, I. et al. (2004). Shear-induced crystallization precursor studies in model polyethylene blends by in-situ rheo-SAXS and rheo-WAXD. *Macromolecules* 37: 4845–4859.

69. MacMillan, S.D., Roberts, K.J., Rossi, A. et al. (2002). In situ small angle x-ray scattering (SAXS) studies of polymorphism with the associated crystallization of cocoa butter fat using shearing conditions. *Crystal Growth and Design* 2: 221–226.

70. Nogales, A., Thornley, S.A., and Mitchell, G.R. (2004). Shear cell for in situ WAXS, SAXS, and SANS experiments on polymer melts under flow fields. *Journal of Macromolecular Science* B43: 1161–1170.

71. Hadziioannou, G., Mathis, A., and Skoulios, A. (1979). Obtention de «monocristaux» de copolymeres trisequences styrene/isoprene/styrene par cisaillement plan. *Colloid and Polymer Science* 257: 136–139.

72. Suehiro, S., Saijo, K., Ohta, Y. et al. (1986). Time-resolved detection of x-ray-scattering for studies of relaxation phenomena. *Analytica Chimica Acta* 189: 41–56.

73. Koppi, K.A., Tirrell, M., Bates, F.S. et al. (1992). Lamellae orientation in dynamically sheared diblock copolymer melts. *Journal of Physics (France) II* 2: 1941–1959.

74. Tepe, T., Schulz, M.F., Zhao, J. et al. (1995). Variable shear-induced orientation of a diblock copolymer hexagonal phase. *Macromolecules* 28: 3008–3011.

75. McConnell, G.A. and Gast, A.P. (1997). Melting of ordered arrays and shape transitions in highly concentrated diblock copolymer solutions. *Macromolecules* 30: 435–444.

76. Langela, M., Wiesner, U., Spiess, H.W., and Wilhelm, M. (2002). Microphase reorientation in block copolymer melts as detected via FT rheology and 2D SAXS. *Macromolecules* 35: 3198–3204.

77. Stangler, S. and Abetz, V. (2003). Orientation behavior of AB and ABC block copolymers under large amplitude oscillatory shear flow. *Rheologica Acta* 42: 569–577.

78. Polushkin, E., van Ekenstein, G.A., Ikkala, O., and ten Brinke, G. (2004). A modified rheometer for in-situ radial and tangential SAXS studies on shear-induced alignment. *Rheologica Acta* 43: 364–372.

79. Polushkin, E., Bondzic, S., de Wit, J. et al. (2005). In-situ SAXS study on the alignment of ordered systems of comb-shaped supramolecules: a shear-induced cylinder-to-cylinder transition. *Macromolecules* 38: 1804–1813.

80. Sota, N., Saijo, K., Hasegawa, H. et al. (2013). Directed self-assembly of block copolymers into twin BCC-sphere: phase transition process from aligned hex-cylinder to BCC-sphere induced by a temperature jump between the two equilibrium phases. *Macromolecules* 46: 2298–2316.

81. Impéror-Clerc, M., Hamley, I.W., and Davidson, P. (2001). Fast and easy flow-alignment technique of lyotropic liquid-crystalline hexagonal phases of block copolymers and surfactants. *Macromolecules* 34: 3503–3506.

82. Castelletto, V. and Hamley, I.W. (2006). Capillary flow behaviour of worm-like micelles studied by small angle X-ray scattering and small angle light scattering. *Polymers for Advanced Technologies* 17: 137–144.

83. Bharati, A., Hudson, S.D., and Weigandt, K.M. (2019). Poiseuille and extensional flow small-angle scattering for developing structure-rheology relationships in soft matter systems. *Current Opinion in Colloid and Interface Science* 42: 137–146.

84. Cloke, V.M., Higgins, J.S., Phoon, C.L. et al. (1996). Poiseuille geometry shear flow apparatus for small-angle scattering experiments. *Review of Scientific Instruments* 67: 3158–3163.

85. Hamley, I.W. (2001). Structure and flow behaviour of block copolymers. *Journal of Physics. Condensed Matter* 13: R643–R671.

86. Hamley, I.W. and Castelletto, V. (2004). Small-angle scattering of block copolymers in the melt, solution and crystal states. *Progress in Polymer Science* 29: 909–948.

87. Kisilak, M., Anderson, H., Babcock, N.S. et al. (2001). An x-ray extensional flow cell. *Review of Scientific Instruments* 72: 4305–4307.

88. Penfold, J., Staples, E., Tucker, I. et al. (2006). Elongational flow induced ordering in surfactant micelles and mesophases. *Journal of Physical Chemistry. B* 110: 1073–1082.

89. Daniel, C., Hamley, I.W., and Mortensen, K. (2000). Effect of planar extension on the structure and mechanical properties of polystyrene-poly(ethylene-*co*-butylene)-poly(styrene) triblock copolymers. *Polymer* 41: 9239–9247.

90. Stasiak, J., Squires, A.M., Castelletto, V. et al. (2009). Effect of stretching on the structure of cylinder- and sphere-forming styrene-isoprene-styrene block copolymers. *Macromolecules* 42: 5256–5265.

91. Stasiak, J., Brubert, J., Serrani, M. et al. (2015). Structural changes of block copolymers with bi-modal orientation under fast cyclical stretching as observed by synchrotron SAXS. *Soft Matter* 11: 3271–3278.

92. Mencke, A., Cheng, A.C., and Caffrey, M. (1993). A simple apparatus for time-resolved x-ray-diffraction biostructure studies using static and oscillating pressures and pressure jumps. *Review of Scientific Instruments* 64: 383–389.

93. Lorenzen, M., Riekel, C., Eichler, A., and Haussermann, D. (1993). A high-pressure-cell for small-angle x-ray-scattering. *Journal de Physique IV* 3: 487–490.

94. Tien, N.D., Sasaki, S., and Sakurai, S. (2016). Influence of high pressure on higher-order structures of poly(oxyethylene) in its blend with poly(d,l-lactide). *Polymer Bulletin* 73: 399–408.

95. Erbes, J., Winter, R., and Rapp, G. (1996). Rate of phase transformations between mesophases of the 1:2 lecithin fatty acid mixtures DMPC/MA and DPPC/PA – a time-resolved synchrotron X-ray diffraction study. *Berichte der Bunsengesellschaft für physikalische Chemie* 100: 1713–1722.

96. Pressl, K., Kriechbaum, M., Steinhart, M., and Laggner, P. (1997). High pressure cell for small- and wide-angle x-ray scattering. *Review of Scientific Instruments* 68: 4588–4592.

97. Pollard, M., Russell, T.P., Ruzette, A.V. et al. (1998). The effect of hydrostatic pressure on the lower critical ordering transition in diblock copolymers. *Macromolecules* 31: 6493–6498.

98. Steinhart, M., Kriechbaum, M., Pressl, K. et al. (1999). High-pressure instrument for small- and wide-angle x-ray scattering II. Time-resolved experiments. *Review of Scientific Instruments* 70: 1540–1545.

99. Woenckhaus, J., Kohling, R., Winter, R. et al. (2000). High pressure-jump apparatus for kinetic studies of protein folding reactions using the small-angle synchrotron x-ray scattering technique. *Review of Scientific Instruments* 71: 3895–3899.

100. Nishikawa, Y., Fujisawa, T., Inoko, Y., and Moritoki, M. (2001). Improvement of a high pressure cell with diamond windows for solution X-ray scattering of proteins. *Nuclear Instruments and Methods in Physics Research Section A Accelerators Spectrometers Detectors and Associated Equipment* 467: 1384–1387.

101. Gabke, A., Kraineva, J., Kohling, R., and Winter, R. (2005). Using pressure in combination with x-ray and neutron scattering techniques for studying the structure, stability and phase behaviour of soft condensed matter and biomolecular systems. *Journal of Physics: Condensed Matter* 17: S3077–S3092.

102. Ando, N., Chenevier, P., Novak, M. et al. (2008). High hydrostatic pressure small-angle x-ray scattering cell for protein solution studies featuring diamond windows and disposable sample cells. *Journal of Applied Crystallography* 41: 167–175.

103. McCarthy, N.L.C. and Brooks, N.J. (2016). Using high pressure to modulate lateral structuring in model lipid membranes. In: *Advances in Biomembranes and Lipid Self-Assembly*, vol. 24 (eds. A. Iglic, C.V. Kulkarni and M. Rappolt), 75–89. London: Academic Press Ltd-Elsevier Science Ltd.

104. Brooks, N.J., Gauthe, B., Terrill, N.J. et al. (2010). Automated high pressure cell for pressure jump x-ray diffraction. *Review of Scientific Instruments* 81: 10.

105. Moller, J., Leonardon, J., Gorini, J. et al. (2016). A sub-ms pressure jump setup for time-resolved x-ray scattering. *Review of Scientific Instruments* 87: 8.

106. Winter, R. (2019). Interrogating the structural dynamics and energetics of biomolecular systems with pressure modulation. In: *Annual Review of Biophysics*, vol. 48 (ed. K.A. Dill), 441–463. Palo Alto: Annual Reviews.

107. Pearson, D.S., Holtermann, G., Ellison, P. et al. (2002). A novel pressure-jump apparatus for the microvolume analysis of protein-ligand and protein-protein interactions: its application to nucleotide binding to skeletal-muscle and smooth-muscle myosin subfragment-1. *Biochemical Journal* 366: 643–651.

108. Osaka, N., Takata, S.I., Suzukia, T. et al. (2008). Comparison of heat- and pressure-induced gelation of beta-lactoglobulin aqueous solutions studied by small-angle neutron and dynamic light scattering. *Polymer* 49: 2957–2963.

109. Caffrey, M. (1985). Kinetics and mechanism of the lamellar gel lamellar liquid-crystal and lamellar inverted hexagonal phase-transition in phosphatidylethanolamine – a real-time x-ray-diffraction study using synchrotron radiation. *Biochemistry* 24: 4826–4844.

110. Gruner, S.M. (1987). Time-resolved x-ray-diffraction of biological-materials. *Science* 238: 305–312.

111. Kriechbaum, M., Rapp, G., Hendrix, J., and Laggner, P. (1989). Millisecond time-resolved x-ray-diffraction on liquid-crystalline phase-transitions using infrared-laser T-jump technique and synchrotron radiation. *Review of Scientific Instruments* 60: 2541–2544.

112. Caffrey, M., Magin, R.L., Hummel, B., and Zhang, J. (1990). Kinetics of the lamellar and hexagonal phase-transitions in phosphatidylethanolamine – time-resolved x-ray-diffraction study using a microwave-induced temperature jump. *Biophysical Journal* 58: 21–29.

113. Laggner, P. and Kriechbaum, M. (1991). Phospholipid phase-transitions – kinetics and structural mechanisms. *Chemistry and Physics of Lipids* 57: 121–145.

114. Kenel, C. and Leinenbach, C. (2015). Influence of cooling rate on microstructure formation during rapid solidification of binary TiAl alloys. *Journal of Alloys and Compounds* 637: 242–247.

115. Toft, K.N., Vestergaard, B., Nielsen, S.S. et al. (2008). High-throughput small angle x-ray scattering from proteins in solution using a microfluidic front-end. *Analytical Chemistry* 80: 3648–3654.

116. Ghazal, A., Lafleur, J.P., Mortensen, K. et al. (2016). Recent advances in x-ray compatible microfluidics for applications in soft materials and life sciences. *Lab on a Chip* 16: 4263–4295.

117. Lopez, C.G., Watanabe, T., Adamo, M. et al. (2018). Microfluidic devices for small-angle neutron scattering. *Journal of Applied Crystallography* 51: 570–583.

118. Rodriguez-Ruiz, I., Radajewski, D., Charton, S. et al. (2017). Innovative high-throughput SAXS methodologies based on photonic lab-on-a-chip sensors: application to macromolecular studies. *Sensors* 17: 12.

119. Beuvier, T., Panduro, E.A.C., Kwasniewski, P. et al. (2015). Implementation of in situ SAXS/WAXS characterization into silicon/glass microreactors. *Lab on a Chip* 15: 2002–2008.

120. Pham, N., Radajewski, D., Round, A. et al. (2017). Coupling high throughput microfluidics and small-angle X-ray scattering to study protein crystallization from solution. *Analytical Chemistry* 89: 2282–2287.

121. Narayanan, T. (2009). High brilliance small-angle x-ray scattering applied to soft matter. *Current Opinion in Colloid & Interface Science* 14: 409–415.

122. Harford, J. and Squire, J. (1997). Time-resolved diffraction studies of muscle using synchrotron radiation. *Reports on Progress in Physics* 60: 1723–1787.

123. Reconditi, M. (2006). Recent improvements in small angle x-ray diffraction for the study of muscle physiology. *Reports on Progress in Physics* 69: 2709–2759.

124. Dickinson, M., Farman, G., Frye, M. et al. (2005). Molecular dynamics of cyclically contracting insect flight muscle in vivo. *Nature* 433: 330–333.

125. Gopinadhan, M., Majewski, P.W., Choo, Y., and Osuji, C.O. (2013). Order-disorder transition and alignment dynamics of a block copolymer under high magnetic fields by in situ x-ray scattering. *Physical Review Letters* 110: 5.

126. McCulloch, B., Portale, G., Bras, W. et al. (2013). Dynamics of magnetic alignment in rod-coil block copolymers. *Macromolecules* 46: 4462–4471.

127. Rokhlenko, Y., Gopinadhan, M., Osuji, C.O. et al. (2015). Magnetic alignment of block copolymer microdomains by intrinsic chain anisotropy. *Physical Review Letters* 115: 5.

128. Gopinadhan, M., Choo, Y., Kawabata, K. et al. (2017). Controlling orientational order in block copolymers using low-intensity magnetic fields. *Proceedings of the National Academy of Sciences of the United States of America* 114: E9437–E9444.

129. Xu, B., Pinol, R., Nono-Djamen, M. et al. (2009). Self-assembly of liquid crystal block copolymer PEG-b-smectic polymer in pure state and in dilute aqueous solution. *Faraday Discussions* 143: 235–250.

130. Hamley, I.W., Fairclough, J.P.A., King, S.M. et al. (1997). A small angle neutron scattering study of the conformation of a side chain liquid crystal poly(methacrylate) in the smectic C phase. *Liquid Crystals* 22: 679–684.

131. Hamley, I.W., Castelletto, V., Lu, Z.B. et al. (2004). The interplay between smectic ordering and microphase separation in a series of side-group liquid crystal block copolymers. *Macromolecules* 37: 4798–4807.

132. Dong, B., Gwee, L., Salas-de la Cruz, D. et al. (2010). Super proton conductive high-purity Nafion nanofibers. *Nano Letters* 10: 3785–3790.

133. Salas-de la Cruz, D., Denis, J.G., Griffith, M.D. et al. (2012). Environmental chamber for in situ dynamic control of temperature and relative humidity during x-ray scattering. *Review of Scientific Instruments* 83: 7.

134. Chen, X.C., Wong, D.T., Yakovlev, S. et al. (2014). Effect of morphology of Nanoscale hydrated channels on proton conductivity in block copolymer electrolyte membranes. *Nano Letters* 14: 4058–4064.

135. Beers, K.M., Wong, D.T., Jackson, A.J. et al. (2014). Effect of crystallization on proton transport in model polymer electrolyte membranes. *Macromolecules* 47: 4330–4336.

136. Nykaza, J.R., Ye, Y.S., Nelson, R.L. et al. (2016). Polymerized ionic liquid diblock copolymers: impact of water/ion clustering on ion conductivity. *Soft Matter* 12: 1133–1144.

137. Jackson, A., Beers, K.M., Chen, X.C. et al. (2013). Design of a humidity controlled sample stage for simultaneous conductivity and synchrotron x-ray scattering measurements. *Review of Scientific Instruments* 84: 7.

138. Langan, P., Forsyth, V.T., Mahendrasingam, A. et al. (1995). Neutron fiber diffraction studies of DNA hydration. *Physica B* 213: 783–785.

139. Sztucki, M., Narayanan, T., and Beaucage, G. (2007). In situ study of aggregation of soot particles in an acetylene flame by small-angle x-ray scattering. *Journal of Applied Physics* 101: 7.

140. Camenzind, A., Schulz, H., Teleki, A. et al. (2008). Nanostructure evolution: from aggregated to spherical SiO2 particles made in diffusion flames. *European Journal of Inorganic Chemistry*: 911–918.
141. Wyslouzil, B.E., Wilemski, G., Strey, R. et al. (2007). Small angle x-ray scattering measurements probe water nanodroplet evolution under highly non-equilibrium conditions. *Physical Chemistry Chemical Physics* 9: 5353–5358.
142. Kim, Y.J., Wyslouzil, B.E., Wilemski, G. et al. (2004). Isothermal nucleation rates in supersonic nozzles and the properties of small water clusters. *Journal of Physical Chemistry. A* 108: 4365–4377.
143. Landron, C., Hennet, L., Coutures, J.P. et al. (2000). Aerodynamic laser-heated contactless furnace for neutron scattering experiments at elevated temperatures. *Review of Scientific Instruments* 71: 1745–1751.
144. Greaves, G.N., Wilding, M.C., Fearn, S. et al. (2008). Detection of first-order liquid/liquid phase transitions in yttrium oxide-aluminum oxide melts. *Science* 322: 566–570.
145. Leiterer, J., Delissen, F., Emmerling, F. et al. (2008). Structure analysis using acoustically levitated droplets. *Analytical and Bioanalytical Chemistry* 391: 1221–1228.
146. Pfrang, C., Rastogi, K., Cabrera-Martinez, E.R. et al. (2017). Complex three-dimensional self-assembly in proxies for atmospheric aerosols. *Nature Communications* 8: 8.
147. Radnik, J., Bentrup, U., Leiterer, J. et al. (2011). Levitated droplets as model system for spray drying of complex oxides: a simultaneous in situ x-ray diffraction/Raman study. *Chemistry of Materials* 23: 5425–5431.
148. Agthe, M., Plivelic, T.S., Labrador, A. et al. (2016). Following in real time the two-step assembly of nanoparticles into mesocrystals in levitating drops. *Nano Letters* 16: 6838–6843.
149. Polte, J., Ahner, T.T., Delissen, F. et al. (2010). Mechanism of gold nanoparticle formation in the classical citrate synthesis method derived from coupled in situ XANES and SAXS evaluation. *Journal of the American Chemical Society* 132: 1296–1301.
150. Ibrahimkutty, S., Wagener, P., Menzel, A. et al. (2012). Nanoparticle formation in a cavitation bubble after pulsed laser ablation in liquid studied with high time resolution small angle x-ray scattering. *Applied Physics Letters* 101: 4.
151. Wagener, P., Ibrahimkutty, S., Menzel, A. et al. (2013). Dynamics of silver nanoparticle formation and agglomeration inside the cavitation bubble after pulsed laser ablation in liquid. *Physical Chemistry Chemical Physics* 15: 3068–3074.
152. Bras, W., Derbushire, G.E., Bogg, D. et al. (1995). Simultaneous studies of reaction kinetics and structure development in polymer processing. *Science* 267: 996–999.
153. Bras, W. and Ryan, A.J. (1998). Sample environments and techniques combined with small angle x-ray scattering. *Advances in Colloid and Interface Science* 75: 1–43.
154. Tashiro, K., Yamamoto, H., Yoshioka, T. et al. (2014). Hierarchical structural change in the stress-induced phase transition of poly(tetramethylene terephthalate) as studied by the simultaneous measurement of FTIR spectra and 2D synchrotron undulator WAXD/SAXS data. *Macromolecules* 47: 2052–2061.
155. Bryant, G.K., Gleeson, H.F., Ryan, A.J. et al. (1998). Raman spectroscopy combined with small angle x-ray scattering and wide angle x-ray scattering as a tool for the study of phase transitions in polymers. *Review of Scientific Instruments* 69: 2114–2117.
156. Chu, B. and Hsiao, B.S. (2001). Small-angle X-ray scattering of polymers. *Chemical Reviews* 101: 1727–1761.
157. Tashiro, K., Kariyo, S., Nishimori, A. et al. (2002). Development of a simultaneous measurement system of X-ray diffraction and Raman spectra: application to structural study of crystal line-phase transitions of chain molecules. *Journal of Polymer Science Part B-Polymer Physics* 40: 495–506.
158. Tashiro, K. and Sasaki, S. (2003). Structural changes in the ordering process of polymers as studied by an organized combination of the various measurement techniques. *Progress in Polymer Science* 28: 451–519.

159. Haas, S., Plivelic, T.S., and Dicko, C. (2014). Combined SAXS/UV-vis/Raman as a diagnostic and structure resolving tool in materials and life sciences applications. *The Journal of Physical Chemistry. B* 118: 2264–2273.

160. Caetano, B.L., Santilli, C.V., Meneau, F. et al. (2011). In situ and simultaneous UV-vis/SAXS and UV-vis/XAFS time-resolved monitoring of ZnO quantum dots formation and growth. *Journal of Physical Chemistry C* 115: 4404–4412.

161. Sanz, A., Nogales, A., Ezquerra, T.A. et al. (2010). Cold crystallization of poly(trimethylene terephthalate) as revealed by simultaneous WAXS, SAXS, and dielectric spectroscopy. *Macromolecules* 43: 671–679.

162. Sanz, A., Ezquerra, T.A., and Nogales, A. (2010). Interplay between amorphous and crystalline domains in semicrystalline polymers by simultaneous SAXS, WAXS and Dielectric Spectroscopy. Synchrotron Radiation in Polymer Science. In: *IOP Conference Series-Materials Science and Engineering*, 14. Bristol: IOP Publishing Ltd.

163. Wurm, A., Lellinger, D., Minakov, A.A. et al. (2014). Crystallization of poly(epsilon-caprolactone)/MWCNT composites: a combined SAXS/WAXS, electrical and thermal conductivity study. *Polymer* 55: 2220–2232.

164. Bras, W., Nikitenko, S., Portale, G. et al. (2010). Combined time-resolved SAXS and x-ray spectroscopy methods. In: *XIV International Conference on Small-Angle Scattering*, Journal of Physics Conference Series, vol. 247 (ed. G. Ungar). Bristol: IOP Publishing Ltd.

165. Zachmann, H.G. and Wutz, C. (1993). Studies of the mechanism of crystallization by means of WAXS and SAXS employing synchrotron radiation. In: *Crystallization of Polymers*, NATO ASI Series (Series C: Mathematical and Physical Sciences), vol. 405 (ed. M. Dosière). Dordrecht: Springer.

166. Wutz, C., Bark, M., Cronauer, J. et al. (1995). Simultaneous measurements of small-angle x-ray-scattering, wide-angle x-ray-scattering, and light-scattering during phase-transitions in polymers. *Review of Scientific Instruments* 66: 1303–1307.

167. Zhou, S.Q., Chu, B., and Dhadwal, H.S. (1998). High pressure fiber optic light scattering spectrometer. *Review of Scientific Instruments* 69: 1955–1960.

# 4

# Applications and Specifics of SAXS

## 4.1  INTRODUCTION

This chapter is devoted to the powerful method of small-angle x-ray scattering (SAXS) and also discusses wide-angle x-ray scattering (WAXS) in the context of simultaneous SAXS/WAXS measurements. SAXS experiments may be performed at synchrotrons as discussed in Section 3.4 or with laboratory instruments as discussed in Section 3.5. There are now many synchrotron SAXS beamlines around the world (Table 3.1) that provide high flux that enables many types of in situ and kinetic measurements. Some examples are presented in this chapter. SAXS measurements on biomolecules (in particular, proteins) in solutions, called BioSAXS, is an important technique in characterizing these macromolecules, which is described in some detail in this chapter. The chapter also provides an overview on selected examples of SAXS studies on soft materials including colloids, polymers, lipids, and self-assembling peptide systems and examples of time-resolved studies, of which there are now a considerable number. Along with BioSAXS users, soft-matter scientists are one of the main communities of SAXS users. SAXS is actually used to investigate many other types of nanomaterial, but lack of space precludes an analysis of all the systems where SAXS has provided valuable information, although some cases such as porous or fractal materials where SANS is also used are discussed in Section 5.11 in

*Small-Angle Scattering: Theory, Instrumentation, Data and Applications,*
First Edition. Ian W. Hamley.
© 2021 John Wiley & Sons Ltd. Published 2021 by John Wiley & Sons Ltd.

the context of SANS. In addition, other examples of SAXS studies using specific sample cells (e.g. studies on materials under shear and in other fields) are discussed in Section 3.12. Unlike SANS, contrast variation in SAXS is not widely employed, although it can be done using anomalous small-angle x-ray scattering (ASAXS) also detailed in this chapter (or with sugar solutions to vary contrast for protein solutions).

This chapter is organized as follows. The production of x-rays in laboratory instruments and at synchrotrons is the subject of Section 4.2. Section 4.3 describes the scattering processes for x-rays, including Thomson scattering, Compton scattering, and polarization-dependent scattering, with a particular focus on the elastic coherent scattering cross-section relevant to small-angle scattering. Atomic scattering factors for x-rays are described in Section 4.4. So-called anomalous SAXS (ASAXS), which exploits the energy-dependent x-ray absorption edges of different elements and which can be used in x-ray contrast variation studies is next discussed in Section 4.5, along with a brief description of other methods to achieve contrast in SAXS. The important subject of solution SAXS from biomacromolecules, so-called BioSAXS, is outlined in some detail in Section 4.6, in the context of individual proteins. Section 4.7 concerns solution SAXS from multidomain and flexible macromolecules. Solution SAXS from multicomponent systems (biomolecular assemblies) is the subject of Section 4.8. Structure factor effects in protein solutions at high concentration are discussed in Section 4.9. Section 4.10 provides a brief resume of SAXS (and WAXS) studies on soft matter and the following sections give a nonexhaustive set of examples. Section 4.11 describes the main features of SAXS/WAXS studies on crystalline polymers. Section 4.12 details the reconstruction of electron density profiles from structure factor peaks, for the particular case of lamellar phases which are commonly observed for self-assembling molecules such as lipids and surfactants. This example is selected since it provides a case where the phase problem that bedevils analysis of x-ray data (Section 1.6.4) can be overcome, based on known constraints. Section 4.13 gives another example of the application of SAXS to the measurement and analysis of the form factor of a variety of self-assembled structures formed by peptides and lipopeptides in aqueous solution. Section 4.14 gives examples of structure factor analysis of colloidal systems, both one-dimensional data used to determine features of the interaction potential and two-dimensional systems showing beautiful oriented colloidal crystal SAXS patterns. Section 4.15 gives a selection of interesting examples of results from SAXS (and SAXS/WAXS) that provide essential insights into the ordering of natural biomaterials. The chapter concludes with Section 4.16 which presents some examples of fast time-resolved SAXS.

## 4.2 PRODUCTION OF X-RAYS

X-rays are produced when electrons are accelerated or decelerated. Lab sources involve bombardment of a metal such as copper (Cu) or molybdenum (Mo) with an electron beam. This may be done in a hot filament or other source (see Section 3.5).

The x-ray spectrum of a metal target contains peaks corresponding to transitions of inner electronic energy levels, K and L levels. Tabulations of these energy levels are available in the online *X-ray Data Booklet* [1]. Figure 4.1 shows the x-ray emission spectrum of copper (at a sufficiently high applied voltage) along with an energy level diagram for the inner levels. The x-ray emission spectrum is not to scale, since the sharp $K_\alpha$ and $K_\beta$ peaks are typically 50–100 times as intense as the Bremsstrahlung [2]. Bremsstrahlung radiation provides a continuous background, it results from the deceleration (the word *bremsen* is German, meaning 'to brake') of electrons by nuclear scattering. The $CuK_\alpha$ radiation is used for many diffraction and SAXS studies, and it can be separated from $CuK_\beta$ radiation using a nickel (Ni) filter, since Ni has an absorption edge at 1.488 Å.

New diffractometer systems have multiple x-ray sources. For example, in addition to a copper tube, a molybdenum tube may also be used. Mo provides harder radiation ($\lambda = 0.7107$ Å), which can also circumvent problems

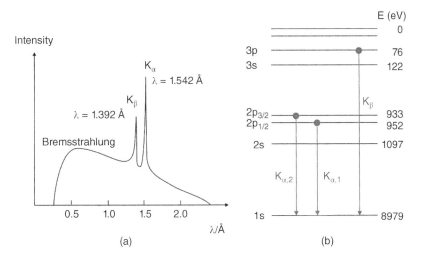

**Figure 4.1** (a) X-ray emission spectrum of copper and (b) associated energy level diagram. The $K_\alpha$ peak is actually a closely spaced doublet with $\lambda(CuK_{\alpha,1}) = 1.54056$ Å and $\lambda(CuK_{\alpha,2}) = 1.54439$ Å. The energy level diagram is shown for convenience with a non-linear energy level scale.

caused by x-ray fluorescence in metallic systems studied with $CuK_\alpha$ x-rays. However, for SAXS instruments, this is not problematic and $CuK_\alpha$ radiation is generally used. Emission energies for K, L, and M lines have been tabulated (see, e.g. the *X-ray Data Booklet*).

X-rays can also be produced at a synchrotron, as described in Section 3.4. The intensity of x-ray sources may be quantified and compared in terms of brilliance, defined as [3]

$$\text{brilliance} = \frac{n}{\text{s.mrad}^2.\text{mm}^2.0.1\%\text{BW}} \qquad (4.1)$$

Here, $n$ refers to number of photons and 0.1% BW refers to the bandwidth with respect to the mean wavelength. This equation normalizes the photon flux per unit area per second, as well as allowing for beam divergence ($\text{mrad}^2$) and bandwidth. The brilliance can be enhanced by keeping the emittance low, this being a measure of the spread of photon position and momentum in a synchrotron beam. Synchrotron beams are characterized by low angular divergence, typically in the range of microradians. The vertical divergence is given by $\Delta\psi = mc^2/E$, where $E$ is the energy of the electrons in the storage ring, and in electronvolt units $mc^2 = 5.11 \times 10^{-4}$ MeV [2, 4].

As shown in Figure 4.2, synchrotron beamlines have a peak brilliance $>10^{15}$ photon $s^{-1}$ $mm^{-2}$ $mrad^{-1}$ $(0.1\%$ BW$)^{-1}$. The corresponding brilliance for fourth generation sources such as the upgraded European Synchrotron Radiation Facility (ESRF) is $>10^{20}$ photon $s^{-1}$ $mm^{-2}$ $mrad^{-1}$ $(0.1\%$ BW$)^{-1}$. The vertical lines in Figure 4.2 show the typical ranges of brightness for different lab sources of x-rays, at peak wavelengths.

Going beyond the flux and brilliance of synchrotrons, x-ray free electron lasers (XFELs) are expected to be the next generation of x-ray source. However, the extreme brilliance (peak brilliance $>10^{30}$ photons $s^{-1}$ $mrad^{-2}$ $mm^{-2}$ $(0.1\%\text{BW})^{-1}$) means that samples of soft and biomaterials studied by SAXS will suffer rapid beam damage although fast measurements should minimize this (an example of an XFEL SAXS study of photolytic breakup of a protein is discussed in Section 4.16). The coherent nature of x-rays produced by free electron lasers also opens up the possibility for new types of coherent x-ray measurements. XFELs are in operation in Europe, United States, Japan, and Korea.

The theory presented in Section 4.3 describes scattering by perfect plane waves (Eq. (4.4)). In reality, this is not the case, due to variability in wavelength and in the direction of the waves. This leads to the concept of coherence, which defines the spatial dimensions over which waves are in phase. Figure 4.3 shows the definition of longitudinal and transverse coherence lengths.

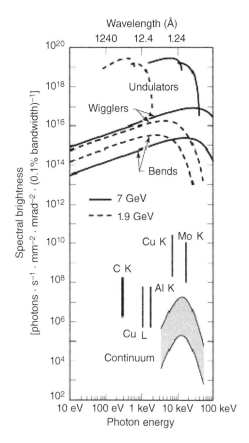

**Figure 4.2** Typical Brilliance of different x-ray sources (data are given for synchrotron rings with two energies). Ranges are given for lower brilliance lab sources of radiation, within which the lower end will be typical of sealed tubes, the middle corresponds to rotating anode sources and the upper end to microfocus sources. *Source:* From Thompson et al. [5]. © 2009, Regents of the University of California.

Considering the longitudinal coherence length, the two waves shown in Figure 4.3a have a difference in wavelength of $\Delta\lambda$ and are in phase at P but move out of phase, coming back into phase at P'. In the distance $2\xi_L$ we have $2\xi_L = N\lambda = (N+1)(\lambda - \Delta\lambda)$, thus $\lambda = (N+1)\Delta\lambda$ and so, using the approximation $N = \lambda/\Delta\lambda$, valid for large $N$ [6],

$$\xi_L = \frac{1}{2}\frac{\lambda^2}{\Delta\lambda} \qquad (4.2)$$

Considering the transverse coherence length, Figure 4.3b shows waves emitted from the two ends of source D, with different directions (angular divergence $\Delta\theta$). The waves are observed at P, a distance $R$ from the source.

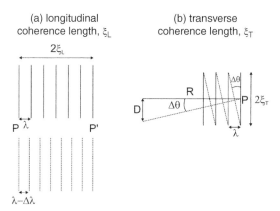

**Figure 4.3** Schematic of plane waves to define (a) longitudinal and (b) transverse coherence lengths.

Geometrically, $\Delta\theta = \lambda/2\xi_T$, but also from the lefthand side of the diagram in Figure 4.3b $\Delta\theta = D/R$, thus [6]

$$\xi_T = \frac{\lambda}{2}\frac{R}{D} \tag{4.3}$$

Synchrotron beams are characterized by a moderate degree of coherence, third-generation sources having typical transverse coherence lengths of 50 μm in the horizontal direction and at least 500 μm vertically at 50–100 m from the source [2]. In fact, modern SR sources are fully coherent in the vertical direction with a source size typically 10–20 μm. The longitudinal coherence length depends on the monochromator [6], for Si(111) normally used at SAXS beamlines, it is ∼1 μm for 1 Å x-rays. Coherence is important for coherent diffraction methods and photon correlation spectroscopy, as well as imaging methods such as phase contrast imaging, coherent diffractive imaging, tomography, and scanning diffraction microscopy. The transverse coherence length should be significantly larger than the size scales probed by SAXS, otherwise, the amplitudes do not sum up as assumed in the theory discussed in Section 1.2, but instead intensities would be additive. The minimum transverse coherence is defined by the collimation setup.

# 4.3   SCATTERING PROCESSES FOR X-RAYS

SAXS relies on elastic scattering of x-rays by electrons. Here we consider the scattering by an electron, and the effect of polarization on the intensity.

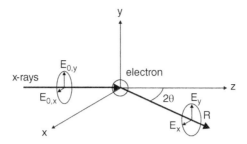

**Figure 4.4**   Thomson scattering process of x-rays by an electron. The x-rays are incident along the $z$ direction and are polarized along the $x$ direction.

Inelastic (Compton) scattering is also defined, although this is not important in SAXS measurements.

Considering elastic scattering of x-rays by electrons, for x-rays polarized in the $x$ direction (Figure 4.4), the incident electric field (strength $E_0$, angular frequency $\omega$) is a vector with magnitude:

$$E_{\text{in}} = E_{x,0}e^{-i\omega t} \tag{4.4}$$

the electric field radiated at a distance $R$ from the electron is (recalling that the wavevector magnitude is $k = 2\pi/\lambda$)

$$\frac{E_{\text{rad}}}{E_{\text{in}}} = -\left(\frac{e^2}{4\pi\varepsilon_0 m_e c^2}\right)\frac{e^{ikR}}{R}\cos 2\theta \tag{4.5}$$

The prefactor is termed the Thomson or classical scattering length of an electron

$$r_e = \left(\frac{e^2}{4\pi\varepsilon_0 m_e c^2}\right) \tag{4.6}$$

This takes the value $r_e = 2.82 \times 10^{-15}$ m. In Eq. (4.6), $e$ is the electron charge, $\varepsilon_0$ is vacuum permittivity, and $m_e$ is the electron mass.

For unpolarised x-rays, the radiated field magnitude is

$$|E_{\text{rad}}|^2 = \left(\frac{r_e^2}{R^2}\right)(E_{y,0}^2 + E_{x,0}^2\cos^2 2\theta) \tag{4.7}$$

This can be averaged to give

$$\langle E_{x,0}^2\rangle = \langle E_{y,0}^2\rangle = \frac{1}{2}\langle E_0^2\rangle \tag{4.8}$$

The differential scattering cross-section is

$$\left(\frac{d\sigma}{d\Omega}\right) = \frac{|E_{\text{rad}}|^2 R^2}{|E_{\text{in}}|^2} \tag{4.9}$$

Thus, for Thomson scattering [6]

$$\left(\frac{d\sigma}{d\Omega}\right) = r_e^2 P \tag{4.10}$$

In the preceding derivation for unpolarised x-rays, the polarization factor $P = \frac{1}{2}(1 + \cos^2 2\theta)$, for vertically polarized x-rays $P = 1$ and for horizontally polarized x-rays $P = \cos^2 2\theta$.

The discussion above concerns elastic scattering. Inelastic scattering processes can also occur, and these lead to so-called Compton scattering, which is a type of incoherent scattering. This leads to a shift in wavelength of the x-rays due to loss of momentum due to collisions with electrons. The resulting wavelength shift is given by [4]

$$\Delta\lambda = \lambda' - \lambda = \frac{h}{mc}(1 - \cos 2\theta) \tag{4.11}$$

The term $h/mc$ (where $h$ is the Planck constant, $m$ is the electron mass, and $c$ is the speed of light) is known as the Compton wavelength of the electron and has the value $\lambda_c = 2.43 \times 10^{-12}$ m. The Compton effect is neglected in all the following discussion, since it is not important in small-angle scattering.

Multiple scattering refers to the scattering of x-rays by more than one electron scattering event. Multiple scattering can be minimised by keeping the sample thickness sufficiently low and is in general not problematic in most small-angle scattering experiments, although it can present serious problems in ultra-small-angle x-ray scattering (USAXS) when the characteristic sizes are large and for high contrast and/or concentrated materials. The sample thickness should be less than the mean free path of the radiation (i.e. the attenuation), which is about 0.01–1 mm for x-rays and 0.5–10 mm for neutrons [7]. However, it can affect the apparent structure factor for concentrated systems of particles or porous materials. Corrections are available for systems that exhibit weak multiple scattering [7].

## 4.4   ATOMIC SCATTERING FACTORS

The atomic scattering factors for x-rays, $f(q)$, are strongly decreasing functions of $q$, as shown by representative plots shown in Figure 4.5.

The functions for a given atom can be described to high accuracy by a sum of exponential functions (and constant) [8]:

$$f(q) = \sum_{i=1}^{4} a_i \exp\left(-\frac{b_i q^2}{(4\pi)^2}\right) + c \tag{4.12}$$

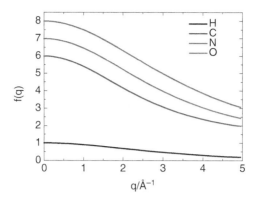

**Figure 4.5**   Atomic scattering factors $f(q)$ for x-rays for the selected atoms indicated.

The limiting value for an atom $j$ is $f_j(q = 0) = n_{e,j}$ where $n_{e,j}$ is the number of electrons for that atom. The parameters $a_i$, $b_i$ and $c$ in Eq. (4.12) are tabulated [8].

## 4.5   ANOMALOUS SAXS AND SAXS CONTRAST VARIATION

In Eq. (4.12), the atomic scattering factors for x-rays refer to values away from absorption edges. Each element has a series of characteristic absorption edges associated with transitions to inner electron energy levels (K and L shells). This leads to an energy- (or equivalently wavelength-) dependent scattering factor [2, 3, 6]:

$$f = f_0 + f'(E) + if''(E) \qquad (4.13)$$

Figure 4.6 shows, as an example, the calculated energy dependence of the coefficients $f'$ and $f''$ for bromine near the K absorption edge at $E = 13.4737$ keV. Electron energies are generally quoted in keV units, which can be converted to joules (J) using $1 \text{ keV} = 1.60218 \times 10^{-16}$ J ($1.60218 \times 10^{-19}$ C is the charge on an electron) or alternatively to wavelength using $\lambda(\text{Å}) = 12.298/E(\text{keV})$. Compilations and plots of energy-dependent atomic scattering factors are available [1, 5, 9].

The strong energy/wavelength dependence of atomic scattering factors near absorption edges underpin the technique of ASAXS and this has been employed as a 'contrast variation' method to locate halide counter ions around both synthetic and natural molecules and macromolecules including surfactants, anionic polymers, and DNA. Figure 4.7 shows representative data obtained for an aqueous solution of a cationic surfactant, tetradecyl

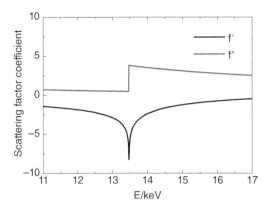

**Figure 4.6** Dependence of atomic scattering factor coefficients for bromine as a function of energy. *Source:* Data from http://skuld.bmsc.washington.edu/scatter/AS_periodic.html.

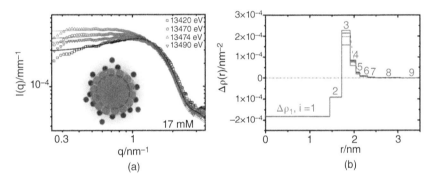

**Figure 4.7** (a) ASAXS data (open symbols) measured for a 17 mM aqueous solution of the surfactant tetradecyl trimethylammonium bromide at different energies (indicated) around the bromine K absorption edge at 13473.7 keV, along with model fits using a multi-shell form factor model (solid lines) (schematic model of micelle and counterions shown inset), (b) Computed scattering density profiles obtained from the model fits (solid lines in part a). *Source:* From Sztucki et al. [10]. © 2012, Springer Nature.

trimethylammonium bromide [10]. The shape of the SAXS profiles in Figure 4.7a changes as a function of the x-ray energy around the bromine absorption edge (Figure 4.6), and the data were fitted (solid lines) to obtain scattering density contrast profiles (Figure 4.7b), which show differences that are due to the distribution of bromine ions, i.e. the distribution of these ions as a function of distance from the micelle core can be calculated more reliably using the additional data from ASAXS measurements.

Resonant x-ray scattering has been extended to the small-angle domain, resonant soft x-ray scattering (RoSoXS) at the carbon K edge has been used to investigate chiral ordering of bonds in liquid crystals. It can be used, for example, to measure crystallographically forbidden reflections for structures such as the bicontinuous gyroid cubic phase with helical molecular packing in the channels or mismatched molecular orientation at the nodes of the network [11].

Contrast variation in SAXS can also be achieved using sugars (such as sucrose or glycerol), or salt to change the electron density of the solvent; indeed, the electron density of a protein is similar to that of a 60% aqueous sucrose solution, and this can be used for contrast matching [2]. Contrast matching with sucrose, glycerol, or salt can be used to contrast match the scattering from some surfactants and polymer latex colloid particles, as well as proteins, but not nucleic acids [12].

## 4.6 BIOSAXS: SOLUTION SAXS FROM BIOMACROMOLECULES, ESPECIALLY PROTEINS

The importance of so-called BioSAXS, which is the use of SAXS to analyse the configuration of biomacromolecules in solution, is attested by the proliferation of dedicated beamlines for this type of experiment at synchrotrons around the world, as well as laboratory instruments. The measurement procedure is described in Section 3.8. Here methods to analyse the data are discussed. First, we will discuss the simplest case, the reconstruction of SAXS (or SANS) profiles from pdb structures, which are usually obtained from single crystal x-ray diffraction, although structures obtained from solution NMR may also be available. Second, we discuss the reconstruction of particle (e.g.) protein shape envelopes using spherical harmonic expansions from which scattering amplitudes can be computed, so that calculated intensity profiles are generated that can be fitted to experimental intensity profiles. Finally, the determination of particle (especially protein) shape using dummy atom models is described. This subject is covered in more detail in excellent reviews [13–20] and dedicated books [2, 21, 22].

### 4.6.1 Determination of Pair Distance Distribution Function

The first analysis of BioSAXS data is usually the determination of the pair distance distribution function $p(r)$ (Section 1.3) and of the radius of gyration from a Guinier analysis (Section 1.4) and of the molar mass from extrapolated $I(0)$ calculations (Section 2.9). These calculations are commonly performed using software available at the beamline via workflow systems such as IspyB [23]. The $p(r)$ profile obtained from an indirect Fourier transform of $I(q)$ can be computed using offline software that performs indirect such as GNOM [24] part of the ATSAS package (Table 2.2). After this, the data may be modelled, fitting either $p(r)$ or $I(q)$ directly as described in the following. The modelling may be low-resolution modelling using bead representations or surface envelope representations of the biomacromolecule shape, or it is possible to calculate small-angle scattering (SAS) profiles from atomistic descriptions of the structure, either from pdb files (crystal or NMR structures) or using co-ordinates from computer simulations.

Software to calculate $p(r)$ relies on indirect Fourier transform methods, first introduced for SAXS data by Glatter [25]. This method has been implemented in generalized indirect Fourier transformation (GIFT) [26–28] and GNOM [24], among other software. In these methods, the pair distance distribution function is represented in the range $r = [0, D_{max}]$ as a sum of orthogonal functions $\phi_j(q_i)$ sampled at $N$ measured $q$ values, $q_i$:

$$p(r) = \sum_{j=1}^{J} c_j \phi_j(q_i) \tag{4.14}$$

The coefficients $c_j$ are obtained by minimization of the functional

$$\Phi = \sum_{i=1}^{N} \left[ \frac{I_{exp}(q_i) - \sum_{j=1}^{J} c_j \psi_j(q_i)}{\sigma(q_i)} \right]^2 + \alpha \int_0^{D_{max}} \left[ \frac{dp}{dr} \right]^2 dr \tag{4.15}$$

Here the $\psi_j(q_i)$ are the Fourier transforms of $\phi_j(r)$ (allowing for smearing if necessary) and $\sigma(q_i)$ are the errors on the measured data. The first term in Eq. (4.15) is a $\chi^2$ term (Section 2.17) relating to the difference between experimental and calculated intensity and the second represents a smoothness constraint, the coefficient $\alpha$ allowing for the balance between the two terms [14, 18]. Indirect Fourier transform methods are discussed in exhaustive detail elsewhere [29].

## 4.6.2    Reconstruction of SAXS Profiles from Known Structures (pdb Files)

Databases such as the protein data bank (http://rcsb.org) provide crystal structures for many proteins. Software that can calculate SAXS or SANS profiles from pdb structures includes CRYSOL and CRYSON (x-ray and neutron versions of essentially the same algorithm, both part of the ATSAS package, Table 2.2) to calculate SAXS and SANS profiles respectively and FoXS (for single state proteins) or MultiFoXS (for multistate proteins), AQUASAXS and AXES (Table 2.2).

The expressions used to compute the SAXS intensity profile differ in CRYSOL and FoXS as discussed in the following. These programmes all rely on use of the Debye formula (Eq. (1.13)) to calculate the scattered intensity from the ensemble of atoms with positions specified in the pdb file. However, the atomic scattering factors in the Debye equation are replaced by terms that allow for the difference in scattering at a particular site with respect to the aqueous solvent (displaced solvent) as well as the hydration layer. The latter is important at the solvated surface of the protein, and description of this is a key part of accurately modelling solution BioSAXS data. This is illustrated in Figure 4.8, which shows an example of calculated contributions to the total scattering intensity using CRYSOL, using bovine serum albumin as a simple example of a protein with several known crystal structures. The scattering factors in the Debye formula calculations in FoXS and CRYSOL are based on treatments of the scattering in terms of allowance for the displaced solvent [30], and (in the case of CRYSOL) representation of the scattering amplitudes in spherical harmonics [31]. They employ different expressions for the scattering factors.

In CRYSOL, the intensity is written in terms of the amplitude in vacuum minus, $A_v(q)$ minus amplitude terms relating to the displaced solvent scattering, $A_s(q)$ and that of the boundary layer (hydrated shell around the

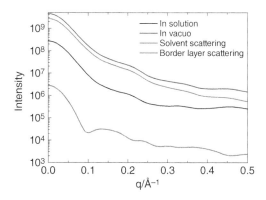

Figure 4.8   Contributions to the intensity for the protein bovine serum albumin, BSA (pdb file 3V03), calculated using CRYSOL.

biomacromolecule), $A_b(q)$ [2, 32]:

$$I(q) = \langle |A_v(q) - \rho_0 A_s(q) + \Delta\rho A_b(q)|^2 \rangle_\Omega \tag{4.16}$$

Here, $\langle\ \rangle_\Omega$ indicates an orientational average and $\rho_0$ and $\rho_b$ are the scattering densities of solvent and boundary layer, respectively, and $\Delta\rho = \rho_b - \rho_0$.

The scattering amplitudes entering in Eq. (1.2) are expanded in spherical harmonics, using the Rayleigh expansion in the form

$$\exp(i\mathbf{q}.\mathbf{r}) = 4\pi \sum_{l=0}^{\infty} \sum_{m=-l}^{l} i^l j_l(qr) Y_{lm}(\hat{\mathbf{q}}) Y_{lm}^*(\hat{\mathbf{r}}) \tag{4.17}$$

where $j_l(qr)$ denotes an $l$th order spherical Bessel function, and the spherical harmonics $Y_{lm}(\hat{\mathbf{x}})$ depend on the orientation angles of unit vectors $\hat{\mathbf{q}}$ and $\hat{\mathbf{r}}$. The boundary layer term in Eq. (4.16) is expressed as a Hankel transform of the scattering density profile of a boundary layer of uniform thickness $\Delta$ with scattering density $\rho_b$. Further details on the derivation of the final series expansion expression for $I(q)$ are available in the original reference [32].

In CRYSOL, the scattering factors of the dummy atoms are written as [30, 32]

$$a_j^s(q) = G(q) V_j \exp\left(-\frac{q^2 V_j^{2/3}}{4\pi}\right) \tag{4.18}$$

Here, $V_j$ is the solvent volume displaced by atom $j$ represented as a Gaussian sphere (values are provided in refs. [30, 32]) The factor $G(q)$ is used to allow for the expansion of the dummy atom:

$$G(q) = \left(\frac{r_0}{r_m}\right)^3 \exp\left[-\frac{\left(\frac{4\pi}{3}\right)^{\frac{3}{2}} q^2 (r_0^2 - r_m^2)}{4\pi}\right] \tag{4.19}$$

Here, $r_m$ is the actual average radius of the atom or group, values of which are tabulated [32], and $r_0$ is the effective value, which may be adjusted during the fitting of the data.

In FoXS the scattering factors in the Debye equation (1.13) are modified from their values in vacuo as in CRYSOL by subtraction of terms due to solvent excluded volume and the border layer [33]:

$$a_j(q) = a_j^v(q) - a_j^s(q) + c_2 s_j a^w(q) \tag{4.20}$$

Here, $a_j^v(q)$ is the scattering factor for atom $j$ in vacuo, which in the case of SAXS is the atomic scattering factor discussed in Section 4.4 and approximated by a sum of exponential functions (Eq. (4.12)). The second term on the righthand side (RHS) of Eq. (4.20) allows for the displaced solvent around dummy atoms, as in Eq. (4.18). In FoXS, $G(q)$ is allowed to vary by 5%, i.e. in the range $0.95 \leq G(q) \leq 1.05$ [33]. The third term on the RHS of Eq. (4.20) allows for scattering from the surface hydration layer, with $s_j$ the solvent-accessible surface area of atom $j$, $a^w(q)$ the scattering factor of water (represented by Eq. (4.18)) and $c_2$ a scaling factor that allows for the difference in the density of the water in the hydration layer and that in the bulk. The solvent-accessible surface area (SASA) may be calculated by using a probe particle or 'rolling ball' (Figure 4.9). There are a number of algorithms to calculate the SASA [34–38]. Usually for computational convenience, H atoms are neglected in the Debye formula calculations.

In MultiFoXS, the user defines a list of flexible residues, and the software then samples the dihedral angles of these residues as part of the least squares fit. AQUASAXS (Table 2.2) is software that breaks down the scattering amplitude in a similar fashion to FoXS, but with different choices for the boundary layer (hydration shell) scattering term in Eq. (4.20) [39]. AXES (Table 2.2), like CRYSOL and FoXS, has a webserver for solution SAXS calculations from pdb files. It performs calculations allowing for explicit water molecules, which can improve fitting compared to CRYSOL [40]. SASTBX can calculate SAXS profiles using the Debye formula, a spherical

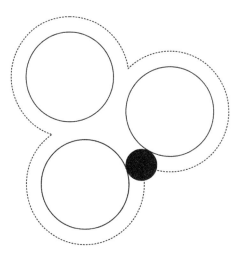

**Figure 4.9** Solvent-accessible surface area (dashed lines) defined using a probe particle. The open circles represent solvent-exposed atoms which are probed with a particle (closed circle) of defined radius. The region between the three circles is inaccessible to the solvent. The calculations are done in three-dimensions during the amplitude factor calculations in CRYSOL and other methods.

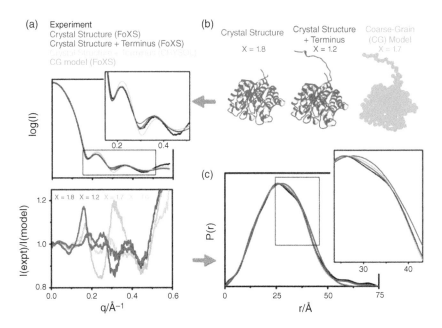

**Figure 4.10** Example to show that accurate modelling of high quality SAXS data requires full atomistic modelling, and provides information on detailed conformation, here showing the effect of allowance for the C-terminal domain. The data are for the cellulose Cel5A catalytic domain. (a) Measured SAXS data compared to intensity profiles shown in (b) calculated using FoXS and CRYSOL software described in the text. The bottom panel in (a) shows the ratio of measured and model intensity. (c) Corresponding pair distance distribution functions, calculated using GNOM (Section 4.6.1). *Source:* From Hammel [43]. © 2012, Springer Nature.

harmonic expansion like CRYSOL or a 3D Zernike expansion [41]. The computational efficiency of these programmes in calculating the SAXS profile from biomolecules have been compared, along with comparison of the formulae used to compute the intensity [42].

It has been shown that high-quality SAXS data is sensitive to atomistic features of protein conformation in solution, as illustrated in Figure 4.10 [43]. This shows that coarse-grained modelling was insufficient to accurately model the SAXS data, whereas atomistic modelling including the C-terminal domain provides a reasonable description (satisfactory $\chi^2$ and low ratio of intensity $I_{experiment}/I_{model}$).

It may seem paradoxical to discuss atomistic-level resolution of a model of SAXS data, when the information content is analysed in terms of the number of independent data points. The naïve resolution from SAXS is estimated as $2\pi/q_{max}$, where $q_{max}$ is the maximum measured $q$ value. The number of independent data points may be estimated according to the number of Shannon channels given by [15]

$$N_s = (q_{max} - q_{min})D_{max}/\pi \qquad (4.21)$$

Here, $D_{max}$ is the maximum distance within the scattering object (Figure 1.3). Using this criterion leads to a typical $N_s \sim 10$–$15$ for many SAXS data sets [15], which is rather low. SAXS data is generally oversampled, in that $\Delta q$ between two adjacent points measured in the scattering curve are is typically less than $\pi/D_{max}$. This fact has been used to argue that the information content of a SAXS curve is higher than that estimated by the number of Shannon channels [15]. Notwithstanding this, it is still desirable to measure down to $q_{min} \leq \pi/D_{max}$. Other estimates of the information content of SAXS data are available [15, 44].

### 4.6.3   Particle Shape Envelope Reconstruction

In a similar way to the representation of the atomic scattering factors as a series expansion in spherical harmonics, the shape envelope of the particle can be expressed in this way. The shape envelope of the particle expressed in spherical coordinates $(r,\omega)$ is expressed as an expansion in spherical harmonics [2, 14, 45, 46]:

$$F(\omega) = \sum_{l=0}^{L} \sum_{m=-l}^{l} f_{lm} Y_{lm}(\omega) \qquad (4.22)$$

where $Y_{lm}(\omega)$ denotes a spherical harmonic and $f_{lm}$ is a multipole coefficient.

The density of the particle is $\rho_c(r) = 1$ within the particle (i.e. for $0 \leq r < F(\omega)$) and $\rho_c(r) = 0$ outside the particle ($r \geq F(\omega)$). This leads to an expression for the intensity [45, 46]

$$I(q) = 2\pi^2 \sum_{l=0}^{\infty} \sum_{m=-l}^{l} |A_{lm}(q)|^2 \qquad (4.23)$$

where the amplitude factors $A_{lm}(q)$ are calculated from the shape coefficients $f_{lm}$ appearing in Eq. (4.22).

The particle shape obtained by dummy atom search optimization methods described in the following section can then be approximated by a shape envelope, as in the software MASSHA (part of ATSAS) [47]. Figure 4.11 shows an example of a shape envelope determined for lysozyme [2]. The ab initio determination of the shape envelope using this method is largely superseded by dummy atom bead modelling methods (Section 4.6.4).

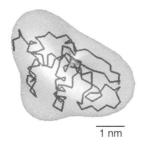

1 nm

**Figure 4.11** Shape envelope of lysozyme (green shaded) and protein backbone (brown line). *Source:* Based on Svergun et al. [2].

### 4.6.4 Dummy Atom Models for SAXS Profile Reconstruction

In these methods, the shape of the particle such as a protein is represented in a highly coarse- grained fashion by a set of beads within a volume containing solvent molecules. The beads are then moved in a simulated annealing algorithm as in the software DAMMIN and DAMMIF (part of ATSAS), and minimising the $\chi^2$ deviation between the calculated and experimental scattering profiles [48]. Figure 4.12 shows an example of a dummy atom model for the protein lysozyme.

## 4.7 SOLUTION SAXS FROM MULTI-DOMAIN AND FLEXIBLE MACROMOLECULES

SAXS probes an ensemble average of the scattering from macromolecules in solution. The preceding discussion in Section 4.6 describes modelling methods applicable to macromolecules such as proteins adopting a single defined conformation. However, many proteins exhibit flexibility, which is important in regulatory mechanisms of biological systems [43]. Proteins may change their conformation in response to stimuli or binding of other molecules or macromolecules. In addition, many proteins have multi-domain structures with flexible linkers between the domains. The accurate modelling of SAXS data from such flexible macromolecules (proteins, protein complexes) requires the calculation of scattering from a set of conformers rather than a single one. There are many approaches to do this, which are outlined in the following. The subject is reviewed in more detail elsewhere [2, 15, 42, 43]. The modelling can be done by direct refinement against SAXS data, or by generation of preassembled conformers (for example by molecular dynamics simulations) and selecting

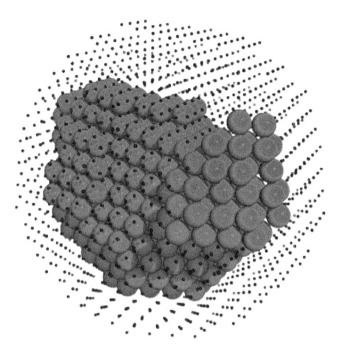

**Figure 4.12**    Protein shape reconstruction using DAMMIN using the simple protein lysozyme as an example (pdb file 6LYZ). The solid red spheres represented the optimized beads for lysozyme, whilst the black dots represent the solvent in the search volume. This is generated in PyMOL from an ATSAS [49] model file.

the best fit model. Distance constraints can be incorporated based on NMR measurements or x-ray crystallography or analytical ultracentrifugation or FRET (Förster resonance energy transfer) or hydrogen–deuterium exchange mass spectrometry, which is sensitive to the conformational states of proteins. Whichever approach is adopted, care needs to be taken not to overfit the data and to avoid samples that are heterogeneous or that contain aggregates [43].

The approach described in Section 4.6 has been generalized to the case of rigid-body modelling of multi-domain proteins in ATSAS routines. MASSHA permits rigid-body modelling of dimers, based on generation of states represented by the three Euler angles and three displacement coordinates between the subunits, and a spherical harmonic representation of the scattering amplitudes [47], as described for unimers in Section 4.6. SASREF may be employed when the subunit structures are known but not the overall structure of the assembly. It performs random rigid-body movements and rotations, using simulated annealing to search for the best fit of the computed SAXS intensity profile to the experimental data [50].

In the simplest case, it calculates the scattering profile from two or more subunits that are interconnected but nonoverlapping. The software is also able to fit multiple sets of experimental intensity profiles – for example, combined SAXS and SANS data with contrast variation series in the SANS data, or data from a series of measurements on mutants, subcomplexes, etc. Constraints on symmetry or from other measurements (for example, NMR) may be included [2]. GLOBSYMM [50] is a further ATSAS package used to model symmetric oligomers via a grid search of possible quaternary structures.

GASBOR is available within ATSAS to model chain-like ensembles using an ab initio method based on dummy residues (Section 4.6.4) [51]. BUNCH applies simulated annealing methods to model SAXS data from multidomain proteins, allowing for multiple data sets from series of deletion mutants [50]. It can be used to model complexes consisting of several subunits, when not all subunit structures are known, and can reconstruct missing loops. The characteristics of these ATSAS routines are summarized in Table 4.1.

The typical process for analysis of solution SAXS data from macro-molecules such as proteins using ATSAS routines is shown in Figure 4.13 [15]. It should be noted that these routines can give ambiguous (nonunique) fits to SAXS data and proposed models should be analysed to check that they are physically meaningful at the level of the constituent residues.

Other groups have developed rigid-body modelling approaches to model solution SAXS data from biomacromolecules. Rigid-body SAXS data modelling was applied using MD to sample constrained inter-domain conformations [52]. FoXSDOCK is able to perform a global search of the docking of rigid subunits, then score the fitting of the calculated SAXS profiles for the models generated to the measured data (after an initial filtering based on radius of gyration comparison) [53]. Finally, it can cluster the remaining models and then perform conformational refinement against the experimental data [53, 54]. Figure 4.14 shows an example of a SAXS profile calculated using FoXSDock for a benchmark protein complex from the protein database [53]. MultiFoXS is a version of FoXS that allows for multistate modelling of conformationally heterogeneous proteins in solution, based on conformational sampling of user-specified flexible residues [54]. Around 10 000 conformers are generated using a Rapidly exploring Random Trees (RRTs) algorithm, and the SAXS profiles (calculated as in FoXS, Section 4.6.2) may be compared to measured data.

Other groups have described the use of Monte Carlo simulations [56, 57], rigid body simulated annealing (with constraints from NMR and with additional SANS data modelled [58]), or minimal molecular dynamics [59, 60] to explore the conformational space (dihedral angles) of connecting regions

**Table 4.1** Characteristic of ATSAS algorithms for rigid body modelling of macromolecular complexes [50].

| Routine characteristic | GLOBSYMM | SASREF | BUNCH |
| --- | --- | --- | --- |
| Structures | Homo- and hetero-dimers | Symmetric oligomers with one monomer per asymmetric part | Multidomain proteins, complexes of subunits with missing fragments |
| Multiple data set fitting | No | Yes | Yes |
| Maximum number of independent rigid bodies | 1 | 10 | 10 |
| Symmetry | P2–P6, P222–P62 | P1–P6, P222–P62 | P1-P6, P222–P62 |
| Minimization method | Global grid search | Simulated annealing | Simulated annealing |
| Constraints | Symmetry, interconnectivity | Symmetry | Symmetry, interconnectivity |
| Restraints | Steric clashes, pair contacts | Interconnectivity, steric clashes, pair contacts | Compactness, steric clashes and bond/dihedral angles in dummy residue loops |
| Number of target function evaluations/CPU, min | 15 000/3 | $1.2 \times 10^5/50$ | $3 \times 10^5/80$ |

*Source:* Adapted from Petoukhov and Svergun [50].

between subunits in multi-domain proteins. The software BILBOMD allows modelling using a minimal MD routine to perform conformational sampling, along with FoXS or CRYSOL to compute the SAXS profile and ensemble analysis using a minimal ensemble search (MES) genetic algorithm [55], Figure 4.15 shows an example of data fitting using this method, comparing the best single conformer fit with that from MES.

Coarse-graining methods have also been used in conformational sampling to model flexible proteins for SAXS data fitting in a basis-set supported reconstruction method [61], or when combining SAXS data

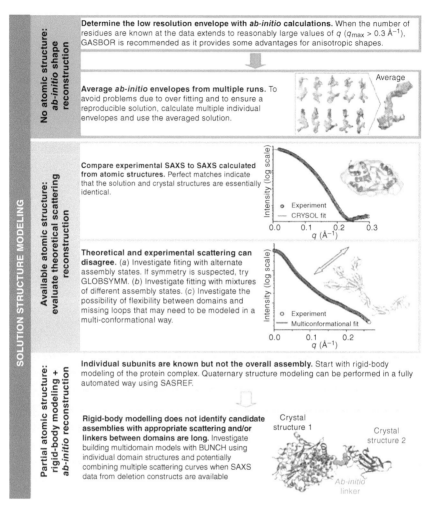

Figure 4.13   Flow chart for protein solution structure SAXS data modelling using ATSAS routines discussed in the text. *Source:* Modified from Putnam et al. [15].

with coarse-grained replica-exchange Monte Carlo (REMC) simulations, using a maximum entropy approach [57]. It should be considered however that, as illustrated in Figure 4.10, coarse-grained models are often not sufficient to fully describe the SAXS intensity profile, atomistic models being required [43].

As an alternative to rigid body methods, approaches to the modelling of SAXS data based on addition of missing fragments (especially loops connecting subunits) have been developed. These fragments may be missing from crystallography or NMR structural information and/or may result

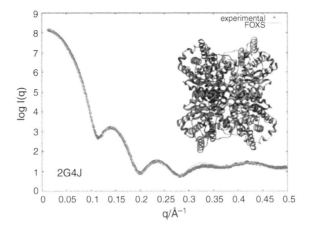

**Figure 4.14** Comparison of measured SAXS profile for glucose isomerase (pdb 2G4J) with a profile calculated using FoXSDock. *Source:* From Schneidman-Duhovny et al. [53].

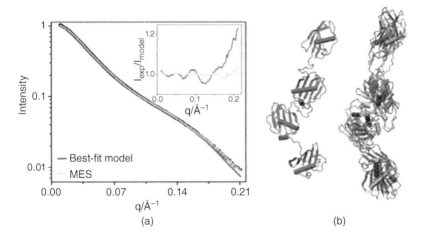

**Figure 4.15** Fitting of SAXS data for extracellular adherence protein (EAP). (a) Black symbols: experimental data, red line: the best-fit single conformer model, green line: minimal ensemble search fit, inset: ratio of experimental/model intensity, (b) red structure: best single conformer model, blue structures: superposition of a mixture of five conformers. *Source:* From Pelikan et al.[55].

from intrinsic macromolecular flexibility or conformational heterogeneity. These missing fragments can be represented as beads or chains of dummy residues, which are subjected to rigid body movements to fit the data. This modelling can be done within ATSAS using the routine

BUNCH discussed above (see also Table 4.1). The rigid body motions are constrained by restrictions on allowed bond and dihedral angles (with reference to Ramachandran plots [2]). BUNCH is designed for modelling of single-chain proteins or symmetric assemblies with one chain per subunit. The ATSAS routine CORAL offers a more general approach that allows components of multiple domains to be moved, as in SASREF [62]. A library of pre-generated self-avoiding random loops is used to generate the conformations, and simulated annealing methods are employed to model the SAXS data using a dummy residue method [2].

## 4.8    SOLUTION SAXS FROM MULTI-COMPONENT SYSTEMS – BIOMOLECULAR ASSEMBLIES

The process of assembly of aggregate structures of biomolecules such as protein amyloid or virus capsids can be analysed by singular value decomposition (SVD). SVD is a matrix method to decompose the components that contribute to a spectrum (or in this case set of SAXS data). The decomposition provides information on the number of components and the basis sets (SAXS profiles) of those components. SVD analysis is included within PRIMUS [63], part of the ATSAS software package (Table 2.2). The program OLIGOMER within ATSAS may then be used to calculate the total intensity from linear combinations of the scattering from individual components, using computed volume fractions [63].

In one example, SVD has been used in the analysis of the amyloid formation kinetics of insulin [64]. The SAXS patterns were interpreted as a superposition of those from monomers, oligomers, and mature fibrils. Using DAMMIN (Section 4.6.4) and DAMAVER (which aligns and averages DAMMIN models [65]) in ATSAS, low-resolution models of protofilaments, protofibrils, and fibrils were reconstructed. This study revealed a helical structure for the oligomer, which acts as the nucleating unit for fibril elongation [64]. A similar study was performed to monitor the fibrillization of the Parkinson's disease related protein α-synuclein by time-resolved SAXS and to determine the structure of the oligomer and elongating fibrils [66].

An additional constraint, the Shannon entropy, has been introduced into the analysis of the SAXS data from a capsid system undergoing component assembly [67]. Writing the total intensity as

$$I_{\text{model}}(q) = \sum_n p_n I_n(q) \tag{4.24}$$

where $p_n$ is the probability of an intermediate (partially assembled) structure and $I_n$ is the associated SAXS intensity, the associated Shannon entropy is

$$S = -K \sum_n p_n ln p_n \tag{4.25}$$

where $K$ is a positive constant. Further details of the numerical analysis employed in the maximum entropy calculation are provided elsewhere [67].

## 4.9 PROTEIN STRUCTURE FACTOR SAXS

The preceding Sections 4.6–4.8 have described the analysis of protein conformation based on solution SAXS, assuming dilute conditions such that structure factor effects can be neglected, as required when measuring form factor features.

In certain cases – for example, in studies of protein aggregation and gelation – it is of interest to study the structure factor in more concentrated solution. As discussed in Section 1.5, the structure factor can be obtained from measured SAXS intensity data by division by the form factor (cf. Figure 1.5 for the case of spherical particles). This is equivalent to dividing the measured intensity in a concentrated solution where structure factor effects are significant by the measured SAXS profile in a sufficiently dilute solution that corresponds to purely form factor scattering. Figure 4.16a shows examples of intensity profiles calculated in this way for the protein bovine serum albumin (BSA) at different concentrations, as indicated [68].

Measurements were also performed at different salt (NaCl) concentrations to examine charge screening effects. The data were fitted using a structure factor calculated from a screened Coulomb potential at low ionic strength, a hard sphere structure factor (Section 1.6.1) at moderate ionic strength, and a square well potential structure factor at high ionic strength. These potentials are shown in Figure 4.16b. The structure factor is the Fourier transform of the radial (pair) distribution function $g(r)$ (Eq. (1.32)), which, in turn, is related to the potential energy $U(r)$ via closure relationships such as the Percus-Yevick relationship [69, 70]:

$$g(r) = \exp\left[-\frac{U(r)}{k_B T}\right] [1 + e(r)] \tag{4.26}$$

where $e(r)$ is the indirect correlation function. At infinite dilution

$$g(r) = \exp\left[-\frac{U(r)}{k_B T}\right] \tag{4.27}$$

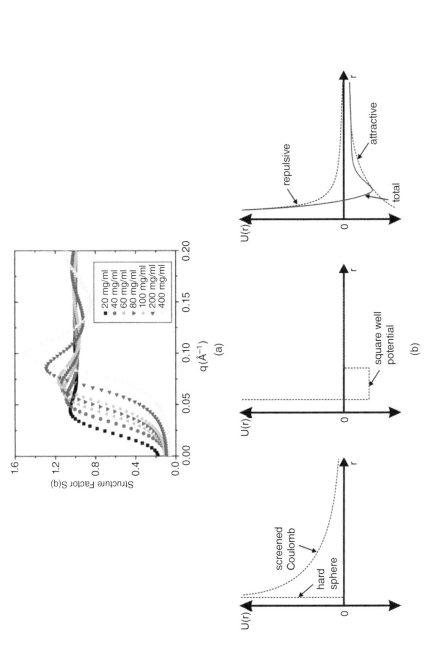

**Figure 4.16** (a) Structure factors at the concentrations indicated obtained from SAXS measurements on the protein bovine serum albumin. *Source:* From Zhang et al. [68]. © 2007, American Chemical Society. (b) Schematic of repulsive potentials, a simple square well potential with an attractive minimum as well as short range repulsion, and addition of repulsive and attractive potentials to give a typical potential.

Closure relationships such as the Percus-Yevick equation and other aspects of liquid state theory are discussed in Section 1.6.7 and elsewhere [2, 70].

The determined structure factor can be used to obtain the second virial coefficient, which provides information on pairwise interactions between protein molecules, in particular the relative strength of short-range attractive interactions that will influence the interaction potential – for example, locally screening repulsive interactions. The second virial coefficient may be obtained from a Zimm plot, Eq. (2.25) in the limit of the structure factor at $q = 0$ [2]:

$$\frac{1}{S(c, q = 0)} = 1 + 2A_2 c \qquad (4.28)$$

Structure factor measurements lead to important insights into liquid–liquid phase separation, crystallization and gelation, which are observed for concentrated solutions of proteins under appropriate conditions. Under certain conditions, local attractive interactions can lead to equilibrium cluster formation, as shown by small-angle scattering data, which contains an additional peak due to the inter-cluster spacing as well as a peak due to the inter-protein spacing within the cluster, as revealed by SANS [71]. Other studies on colloidal systems such as poly(methyl methacrylate) (PMMA) latex particles or lysozyme protein with short-range attraction and long-range repulsion have been reviewed [72].

Aggregation into larger-scale protein complexes can be studied by USAXS or USANS (Section 5.10) or Spin Echo SANS (SESANS) (Section 5.13). Examples include USANS studies of gelation in milk [73], investigation of casein micelle (phosphoprotein aggregates found in milk) structure by USAXS [74, 75], and particle sizing in dairy products investigated by SESANS, such as milk and yogurt with particles with radii up to 5 μm [73, 76].

## 4.10   SAXS (AND WAXS) STUDIES OF SOFT MATTER SYSTEMS

SAXS has been extensively used to investigate soft materials, which include polymers, colloids, liquid crystals, surfactants, and certain biomolecules (lipids, peptides, and nucleic acids and their mixtures) [77–80]. Aspects of the use of small-angle scattering to analyse the conformation of polymer chains and polymer blends are discussed in Sections 1.8 and 5.8. Section 3.12 gives examples of the study of soft materials in various applied external fields, such as shear fields (so-called rheo-SAXS) and kinetic studies of self-assembly under stopped-flow mixing conditions.

SAXS on soft materials may be performed on liquids such as surfactant or colloid solutions or melts of polymers or thermotropic liquid crystal phases. SAXS has also been widely used to probe the structure of gels and other soft solids. SAXS studies of soft materials such as colloids, polymers, and surfactants have been the subject of many reviews and books [3, 4, 81–86]. In the following sections, additional selected examples are discussed.

## 4.11  SAXS AND WAXS FROM SEMICRYSTALLINE POLYMERS

Semicrystalline polymers are characterized by a hierarchical structure [77], which ranges from the sub-nm unit cell structure of the crystalline polymer chains (typically extended along the *c*- axis and often with helical order) up to mm–mm scale spherulite structures, which can be seen by optical microscopy or even with the naked eye, via the characteristic 1–10 nm spacing of the lamellae that characterize the structure of semicrystalline polymers (Figure 4.17).

Ordering of the polymer chains within the unit cell and the development of crystallinity can be monitored by WAXS and time-resolved WAXS respectively. Simultaneous SAXS permits the lamellar ordering to be probed. Again,

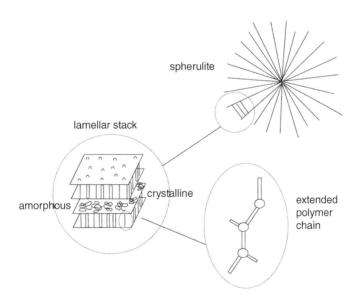

**Figure 4.17**    Schematic showing hierarchical order of semicrystalline polymers.

this may be done in real time following a temperature quench from high temperature to a temperature below the crystallization temperature.

Combined SAXS/WAXS can be performed with some laboratory instruments and some offer simultaneous SAXS/WAXS with two detectors. Simultaneous SAXS/WAXS measurements using separate detectors can be performed on beamlines at a number of synchrotrons internationally (see Table 3.1).

Due to the expected $q^{-2}$ scattering from a lamellar system (Eq. (1.86)), SAXS data from semicrystalline polymers may be presented as so-called Lorentz-corrected plots, i.e. $q^2 I(q)$. Figure 4.18 shows such a plot for a polyethylene-containing polymer. These data confirm the semicrystalline nature of PE because the second Bragg peak intensity is higher than that expected for a stack of crystal lamellae, the additional scattering from amorphous polymer can be represented as a contribution from a broad peak centred near $q = 0.05\ \text{Å}^{-1}$ as shown in Figure 4.18. This peak is not a real peak but is maximum emphasized in a plot of $Iq^2$ (i.e. the same form as a Kratky plot discussed in Section 2.14).

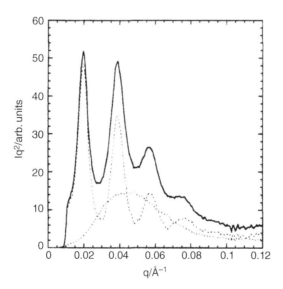

**Figure 4.18** Lorentz-corrected SAXS data for a PE-containing polymer (a polyethylene-*b*-polyethylethylene diblock copolymer) containing semicrystalline PE. The measured data (solid line) can be decomposed as a sum of scattering from a lamellar phase (dashed line with series of Bragg peaks) and a broad peak (dotted line) centred near $q = 0.05\ \text{Å}^{-1}$ (due to amorphous PE). *Source:* From Ryan et al. [87]. © 1995, American Chemical Society.

The invariant is a measure of the total small-angle scattering of a material and in the case of a semicrystalline polymer, Eq. (2.27) is applicable, with $\Phi = \phi_c$, the volume fraction of crystallized polymer.

The lamellar spacing (long period) and thickness of the crystalline layer can be obtained from the one-dimensional correlation function defined as [83, 88, 89]

$$\gamma_1(r) = \frac{\int_0^\infty I(q)q^2 \cos(qr)dq}{\int_0^\infty I(q)q^2 dq} \tag{4.29}$$

Since it is not possible in practice to perform the integration over $q$ extending to infinity, in practice an extrapolation of the intensity to this limit is performed using a Porod law (Section 2.13) function [83, 87].

Figure 4.19 shows a schematic of a correlation function along with the crystal layer thickness $d_c$, which is obtained by extrapolation of the initial linear decay and the long period $L$, which is given by the position of the first maximum in the correlation function. Additional information on the sharpness of the crystal layer interface may be obtained from the decay of the correlation function [4, 89, 90] and it will have a more complex shape for multicomponent systems such as semicrystalline block copolymers which contain a noncrystallizable block as well as a semicrystalline one (see for example Ref. [87]).

Similar to the SAXS data, WAXS data from a crystallized polymer contains a contribution from crystalline and amorphous material as shown in

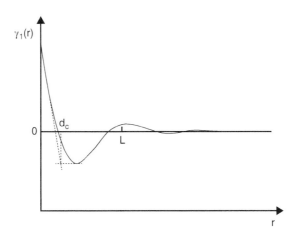

**Figure 4.19** Schematic of a SAXS 1D-correlation function for a semicrystalline polymer showing the extrapolation to obtain the crystal layer thickness $d_c$ and the long period (lamellar spacing) $L$.

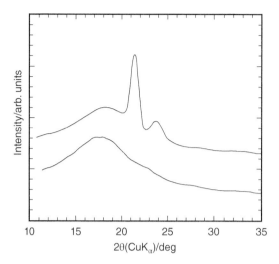

**Figure 4.20** WAXS data for a PE-containing polymer (a polyethylene-*b*-polyethylethylene) diblock copolymer containing semicrystalline PE. Top: data for a crystallized sample, bottom: data for a sample in the melt. *Source:* From Ryan et al. [87]. © 1995, American Chemical Society.

the example in Figure 4.20. The WAXS data for the PE-containing polymer at high temperature has a broad halo centred on $2\theta(CuK_\alpha) = 17.5°$ (lower profile in Figure 4.20). This is due to the amorphous polymer (the feature is due to the preferred polymer chain–chain-packing distance). This feature is still present when the polymer crystallizes (upper WAXS profile in Figure 4.20) but now is present as a shoulder, supplemented by the first two Bragg peaks for crystalline polyethylene (PE).

The areas under the features due to amorphous (broad 'background' peak) and the Bragg peaks can be used to estimate the degree of crystallinity from the relative integrated areas of the peaks obtained by a suitable peak-fitting method:

$$\phi_c = \frac{I_c}{I_c + I_a} \tag{4.30}$$

Here $\phi_c$ is the relative degree of crystallinity and $I_c$ and $I_a$ are, respectively, the integrated intensities of a selected crystal peak and the amorphous broad peak. This provides an indication of the relative degree of crystallinity and can be used to monitor changes in this quantity as the sample crystallizes or melts. Absolute values of the degree of crystallinity cannot be obtained from WAXS data [83]. In a WAXS pattern, the average crystallite size, $l_c$, is related to the width of a Bragg peak at an angle $\theta_B$ (resolution-corrected full width at half maximum, $\Delta\theta_B$) via the Scherrer equation [85, 91]

$$l_c = \frac{K\lambda}{\Delta\theta_B \cos\theta_B} \tag{4.31}$$

Here $K$ is the Scherrer constant, which depends on the shape of the crystallite, $K = 0.9$–$1.1$ typically, with $K = 0.94$ for a cube-shaped crystal [3, 85, 91].

Quantitative analysis of WAXS data is difficult, requiring allowance for the orientation of the detector, polarisation, and in some cases Lorentz correction. The latter allows for the greater contribution of crystallites to the scattering at low angles [3, 83].

Time-resolved SAXS/WAXS is an excellent method to monitor the kinetics of crystallization of polymers, which takes place on the timescale of minutes to hours, or less, down to the millisecond timescale, during processing such as extrusion, for example. The degree of crystallinity increases with time as a sigmoidal (S-shaped) curve, which is described by an Avrami equation of the form $\phi_c = 1\text{-}\exp(-kt^n)$ where $k$ is a rate constant and $n$ is an exponent that can be used to infer information on the dimensionality/mode of growth [77, 89].

## 4.12   LIPID PHASES: ELECTRON DENSITY PROFILE RECONSTRUCTION

Lipids self-assemble into a number of lamellar phases in aqueous solution (as well as hexagonal and cubic phases). The structure factors of a lamellar phase under different levels of approximation are discussed in Section 1.6.1. The following describes quantitative SAXS analysis of the intensities of the Bragg reflections from a lamellar structure to determine the electron density profile. This procedure has yielded valuable insights into the ordering of lipid molecules. The analysis described in the following will apply for other lamellar-forming systems also.

The Fourier transform of Eq. (1.51) gives a general expression for the electron density in terms of the amplitude structure factor terms $F_{hkl}$ [4]:

$$\rho(\mathbf{r}) = \frac{1}{V} \sum_{h=-\infty}^{\infty} \sum_{k=-\infty}^{\infty} \sum_{l=-\infty}^{\infty} F_{hkl} \exp[2\pi i(h\mathbf{a}^* + k\mathbf{b}^* + l\mathbf{c}^*).\mathbf{r}] \tag{4.32}$$

However, as mentioned in Section 1.6.4, although the amplitude is the square root of the intensity, $F_{hkl} = (I_{hkl})^{1/2}$ the phase cannot in general be determined. For centrosymmetric (one-dimensional) lamellar structures such as lipid bilayer structures, though, the phase factor is 0 or $\pi$, and the electron density can be written [92–95]:

$$\rho(z) = \sum_{l=1}^{\infty} g_l |F_l| \cos\left(\frac{2\pi l z}{d}\right) \qquad (4.33)$$

Here $g_1$ is a factor that is either + or −, $|F_l|$ is the magnitude of the $l$th structure factor peak, and $d$ is the lamellar spacing. This equation can be used to reconstruct the electron density profile, examples of which are shown for a lipid $L_\alpha$ phase in Figure 4.21. This can be done by taking all possible combinations of the sign of $g_l$ and inspecting the computed $\rho(z)$ profiles and their physical meaningfulness. In addition, constraints on known electron density values may be imposed when the data is on an absolute scale, then the electron density of the aqueous phase region should be around 0.33 electron $\text{Å}^{-3}$, around 0.30 electron $\text{Å}^{-3}$ for the hydrocarbon chain region, and around 0.16 electron $\text{Å}^{-3}$ for the terminal hydrocarbon methyl groups [93, 96, 97]. For a lipid bilayer structure, a strong dip in the relative electron density corresponding to the midpoint of the hydrophobic lipid region is expected along with maxima near the headgroups on each side of the bilayer. This leads to a profile resembling that shown in Figure 4.21c. The other profiles in Figure 4.21 correspond to the indicated sequences of $g_l$ values. In fact, the sequence of the first five phases is known to be − − + − − for many lipids based on prior work [94, 98].

An alternative or supplementary procedure to enhance the constraints on phases is to measure $F_l$ in a hydration series experiments for a series of samples. The function $g_l|F_l|$ may be plotted for different $g_l$ values and should be a smooth function for an optimal solution. This plot can also be prepared for molten lamellar systems such as the polymorphic layer phases of triacylglycerol lipids. These molecules are key components of cocoa butter in chocolate and understanding the molecular packing and its sensitive dependence on temperature is of great interest in terms of the tempering and mouth feel of chocolate. Consequently, many groups have performed SAXS on cocoa butter and chocolate [99–102], although few have determined electron density profiles of different phases. Figure 4.22 shows a computed smooth profile for the triacylglycerol SOS (containing stearyl-oleoyl-stearoyl chains, SOS = 1,3-distearoyl-2-oleoyl-$sn$-glycerol). Instead of hydration series measurements, data for different polymorphs are presented together and can be seen to lie on the same smooth curves. Both lipids and triacylglycerols have quite small lamellar spacings and so are analysed by SAXS with a short camera length, sometimes the scattering is considered (larger $d$-spacing) x-ray diffraction or medium angle diffraction.

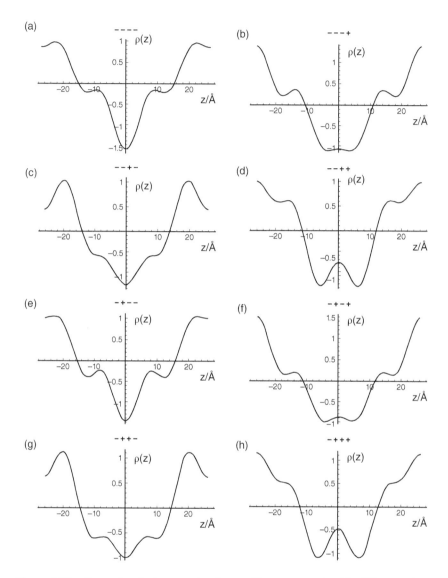

**Figure 4.21** Example of electron density profiles computed with different phase sequences indicated, based on SAXS data for the lipid di-oleoyl phosphatidyl ethanolamine-*sn*-glycerol (DOPE) in a fully hydrated $L_\alpha$ phase at 55 °C. The electron density is plotted on a relative scale. *Source:* Adapted from Harper et al. [96].

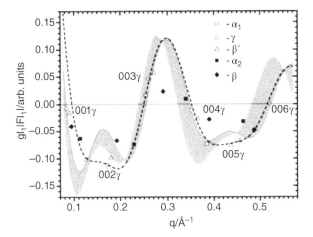

**Figure 4.22** The structure factor term in Eq. (4.33) plotted for different polymorphs (as indicated) for the triacylglycerol SOS, with peaks for the $\gamma$-phase highlighted. The shaded region corresponds to simulated profiles. *Source:* From Mykhaylyk and Hamley [102]. © 2004, American Chemical Society.

The preceding examples concern lipid bilayer structures with symmetric density profiles, for vesicular systems the electron density profile may be asymmetric since the curvature is different for the inner and outer lipid leaflets. This has been incorporated into the modelling of SAXS profiles [103–107]. These models employ different form factor models, including concentric shells, asymmetric electron density slab models, and asymmetric Gaussian functions.

## 4.13 SAXS STUDIES OF PEPTIDE AND LIPOPEPTIDE ASSEMBLIES

Peptides and peptide conjugates including lipopeptides and polymer–peptide conjugates can self-assemble into a variety of different nanostructures. The SAXS form factor from an unaggregated peptide can readily be distinguished from that of peptide assemblies, as shown through the examples presented in Figure 4.23. Figure 4.23b shows the shape of a monomer form factor, which has been fitted with a Debye function for a Gaussian coil (Eq. (1.107)) to represent the unordered conformation. It has a characteristic flat shape at low $q$, curving over at high $q$. The shape of this form factor can be contrasted from those of the assembled structures shown in Figure 4.23.

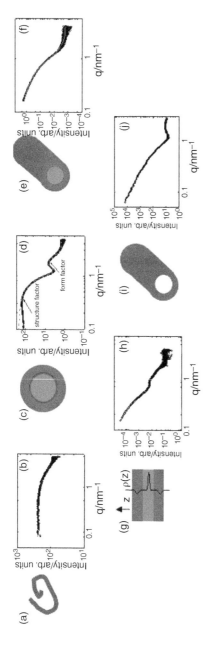

**Figure 4.23** Examples of SAXS data from peptides and lipopeptides in aqueous solution. (a) Gaussian coil representing a monomeric peptide, (b) SAXS data for a 1 wt% aqueous solution of $Pro_6$-Glu$_1$P$_6$E (open symbols) with fit to Gaussian coil form factor (red line). *Source:* From Hamley et al. [108]. (c) sketch of core-shell spherical micelle, (d) SAXS data from a 1 wt% solution of lipopeptide $C_{16}$-CSK$_4$RGDS [$C_{16}$ denotes a hexadecyl chain] with fit to core-shell micelle form factor (dashed red line) and also allowing for structure factor (solid red line) (e) Schematic of core-shell cylinder, (f) SAXS data for a 1 wt% aqueous solution of $A_6R$ (open symbols) with fit to core-shell cylinder form factor (red line). *Source:* From Castelletto et al. [109]. (g) Schematic of a bilayer with superposed electron density profile for a bilayer Gaussian model [110] (three Gaussian representation), with large dip in the lipid interior (blue lamella) and positive relative electron density in the peptide headgroup regions (red domains). (h) SAXS data for an 0.5 wt% aqueous solution of lipopeptide $C_{16}$-YEALRVANEVTLN (open symbols) with Gaussian bilayer form factor fit. *Source:* From Castelletto et al. [111]. (i) Schematic of a nanotube, (j) SAXS data for an 1 wt% aqueous solution of lipopeptide $C_{16}$-KKFFVLK (open symbols) with nanotube (i.e. hollow cylindrical shell) + Gaussian bilayer (to account for electron density cross-section across the nanotube wall) form factor fit. *Source:* From Hamley et al. [112].

SAXS data from lipopeptide/peptide micelles has been fitted using a core–shell (two electron density level) sphere form factor [113–119], an example being shown in Figure 4.23d. This figure also shows the influence of the structure factor, which leads to a broad peak at low $q$. For the spherical micelle case, the structure factor may be the simple, hard-sphere structure factor model (Section 1.6.1), as in the fit in Figure 4.23d.

Amyloid fibrils are formed my misfolded proteins and many model peptides and peptide derivatives that adopt β-sheet secondary structures. A uniform [120] or core-shell [121] cylinder form factor can be used to fit SAXS (or SANS) data from amyloid structures. A core-shell model is required when there is significant contrast difference between the core of the peptide fibril and the exterior, a representative fit is shown in Figure 4.23f. A core–shell form factor model is often required for lipopeptides or polymer–peptide conjugates that form fibrils, as in the case, for example, of PEG-peptides with a PEG corona [121, 122]. For nanotapes formed in particular (but not exclusively) by lipopeptides, a form factor developed for lipid bilayers can be used [123–125]. Figure 4.23h shows an example of data fitted using this form factor. The so-called Gaussian bilayer form factor comprises three Gaussian functions, one of which represents the electron density (in the case of SAXS) of the lipid-chain rich core (negative amplitude), the other two being (positive amplitude) Gaussians representing the electron density of the charged head-groups (Figure 4.23g, cf. Figure 4.21c but with lower resolution) [110]. In some cases for lipopeptide bilayer systems, it is necessary to account for structure factor that can be modelled using the Caillé structure factor [126] (Eq. (1.49)) to represent multi-bilayer ordering [123].

Amphiphilic peptides can assemble into nanotubes, and data have been fitted with corresponding form factors (see, for example, Figure 4.23j) [124, 127, 128]. If the nanotube radius $R \gg t$, where $t$ is the tube wall thickness (e.g. of a bilayer), then the Gaussian bilayer form factor used above can be used to fit the high $q$ part of the small-angle scattering data [112]. Peptide nanotube and other form factors used for self-assembled peptide materials such as fractal form factors have been reviewed [129].

## 4.14 SAXS STUDIES OF THE STRUCTURE FACTOR OF COLLOIDS

Colloids are model systems for understanding interparticle interactions. SAXS studies on charge- and sterically stabilized colloids have yielded insights into the interaction potential via analysis of the structure factor. Figure 4.24 shows an example of structure factor data obtained from

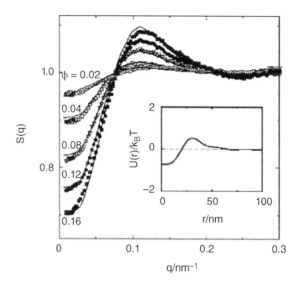

**Figure 4.24** Structure factor obtained from measured SAXS data for laponite colloidal particles at the indicated volume fractions (symbols) along with structure factors calculated using integral equation theory (lines). The inset shows the calculated effective potential $U(r)$. *Source:* From Li et al. [130]. © 2005, Ameircan Physical Society.

aqueous colloidal suspensions of disc-shape laponite mineral particles at several volume fractions [130]. The structure factors were obtained by dividing the intensity measured at finite concentration with the 'form factor' obtained by extrapolation to $\phi \to 0$, where $\phi$ is the volume fraction (this procedure is illustrated in Figure 1.5). The data are compared to structure factors calculated using the Ornstein–Zernike relationship together with the hypernetted chain closure relationship (Section 1.6.7). The latter allows the interaction potential to be determined and, as shown in the inset in Figure 4.24, this comprises an attractive minimum at short distances along with a long-range repulsive contribution. The short-range interaction may be due to local attractive electrostatic interactions between the faces and edges of the discs which bear different charges (at longer ranges electrostatic repulsion between the negatively charged disc faces predominates) [130]. This is proposed to lead to aggregation and gelation observed for laponite at higher concentrations.

In related work, the interaction potential between sterically stabilized (stearoyl functionalized) spherical silica particles was investigated using SAXS to determine the structure factor, which was compared to structure

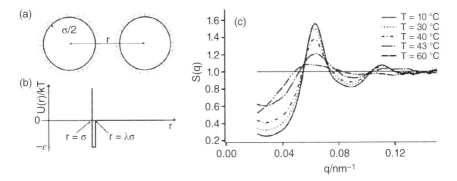

**Figure 4.25** SAXS structure factor analysis for a sterically stabilized silica particle dispersion. (a) schematic of particles showing surface coating of stearoyl chains, (b) Schematic of attractive square-well potential, (c) measured structure factor data at the temperatures indicated for a concentrated solution (volume fraction ~0.12). *Source:* From Pontoni et al. [131]. © 2003, AIP Publishing LLC.

factors computed for an attractive square well potential (Figure 4.16b shows a square well potential) via the Percus-Yevick closure relationship (Section 1.6.7) [130, 131]. Figure 4.25 shows a schematic of the particles, the parameterization for the potential and measured structure factor data as a function of temperature for the system, which shows the development of increasing attractive interactions and kinetic arrest, i.e. 'jamming' (and ultimately a glass transition) of the colloid particles as temperature increases.

SAXS has also been used to probe the orientation of colloidal aggregate liquid crystal and crystal phases. Figure 4.26 shows an example of SAXS data obtained from a colloidal suspension of sterically stabilized gibbsite platelets with added non-adsorbing polymer (polydimethylsiloxane), which causes a short-range depletion interaction [132]. Figure 4.26a,b show the sample (aged after seven months to achieve a steady-state gradient of osmotic pressure down the sample in a capillary). SAXS patterns at the positions indicated in Figure 4.26b are shown in Figure 4.26c–g, along with one-dimensional integrated profiles in Figure 4.26h. The Bragg peaks observed for the columnar phase indicate hexagonal ordering, high-resolution measurements enabled this to be distinguished as hexatic order, i.e. it is characterized by long-range bond orientational order but short-range translational order.

In some cases, microradian resolution of Bragg peaks from colloidal crystals has been possible from synchrotron SAXS measurements [133, 134].

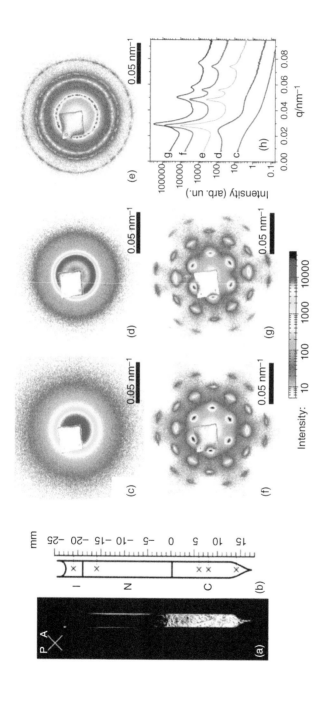

**Figure 4.26** Example of high resolution SAXS data from a suspension of sterically stabilized colloidal gibbsite discs in toluene, in the presence of nonadsorbing polydimethylsiloxane. (a) Polarized optical microscope image showing birefringence in the sample in a capillary, (b) schematic of phases from top: I isotropic, N nematic, C columnar (scale bar in mm); crosses mark positions at which SAXS patterns were measured, (c–g) SAXS patterns including those from spontaneously aligned sample (c) isotropic phase, (d) nematic phase, (e) polydomain columnar phase, (f,g) monodomain columnar phase, (h) integrated one-dimensional SAXS profiles. *Source:* From Petukhov et al. [132]. © 2005, American Physical Society.

# 4.15  SAXS AND SAXS/WAXS STUDIES OF BIOMATERIALS

Biomaterials often comprise hierarchical structures including those with order on the length scales that can be probed using SAXS or SAXS/WAXS. This section provides selected examples of studies on several of the more common structured biomaterials found in animals (skin, bone, hair, collagen, spider silk) and plants (cellulose). Considering other types of animal tissue, some details on SAXS and WAXS studies on muscle structure under electrical stimulation are discussed in Section 3.12.7. This section cannot cover all of the many SAXS studies on other types of biomaterial of which there is a long list.

The first example is collagen, which, in fact, has been used as a calibration for the $q$-scale on older generation synchrotron SAXS beamlines since (properly prepared) it gives a long series of sharp and evenly spaced Bragg reflections from the collagen superstructure. Collagen comes in a variety of forms [135, 136]. The triple helical structure of collagen has a long period of 67 nm in the case of wet rat-tail tendon [137–140]. The third, fifth, ninth, and twelfth orders show enhanced intensity, which facilitates indexing of the pattern. Figure 4.27a gives an example of a SAXS pattern obtained from rat-tail tendon type I collagen, the Bragg reflections being observed parallel to the collagen fibre axis. Figure 4.27b presents a schematic showing the hierarchical order of collagen, extending from the 67 nm periodicity that results from the staggered arrangement of collagen helices to the 0.84 nm periodicity within the collagen helix (collagen is comprised of a repeat of GXY residues where G is glycine and X and Y are other residues [141]).

Another example of a system studied by SAXS/WAXS with hierarchical order is bone, which actually has a structure related to that of collagen since bone is a mineralised tissue based on hydroxyapatite crystals in the presence of calcium and phosphate ions within a collagen matrix. This is clear from Figure 4.28, which shows SAXS data from a bone sample along with a schematic of the order on several length scales in bone. Since bone has an anisotropic structure, the SAXS pattern is also highly anisotropic and Figure 4.28b shows data integrated in radial and axial directions [143]. The SAXS profile in the axial direction shows peaks arising from periodicities within the collagen superstructure. Scanning microfocus SAXS (Section 3.11) has been used to map this orientation across bone samples i.e. scanning the texture [144, 145].

The ordering within many other mineralised biomaterials such as enamel [146, 147], dentin [148, 149], chitin [150], or shells [151, 152] have been investigated by SAXS, as also discussed in reviews [153].

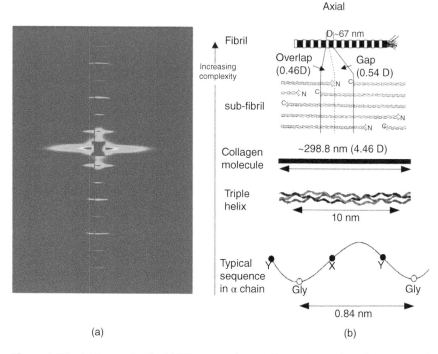

(a)

(b)

**Figure 4.27** (a) Example of a SAXS pattern from collagen (rat-tail tendon, mounted vertically), showing multiple orders of reflection from the 67 nm superstructure. *Source:* From ref. [140]. (b) Hierarchical order within collagen showing characteristic distances [138, 139]. *Source:* From Orgel et al. [138]. © 2001, Elsevier.

A further example of a biomaterial studied by SAXS/WAXS is the stratum corneum in skin [154, 155]. The stratum corneum is a vital structural barrier and it comprise corneocyte cells embedded in lipid bilayers. SAXS on human skin reveals the packing of the lipids in the bilayer, dependent on the hydration state as illustrated in Figure 4.29. The first-order peak corresponds to a spacing 6.5 nm, which is typical of lipid bilayer structures.

Spider silk is a remarkable material, with unique mechanical properties for a lightweight and finely threaded material and has been an interesting model system for in situ SAXS studies (on threads being extended, an example of which is discussed below) as well as conventional measurements. Silk is a semicrystalline material with stacks of crystalline and amorphous domains from the alanine-rich protein [135, 141, 156]. The crystalline domains comprise ordered β-sheet structures. Figure 4.30a shows examples of SAXS profiles from spider dragline silk under different conditions of hydration, stretching, and temperature [157]. The Bragg peaks (first and

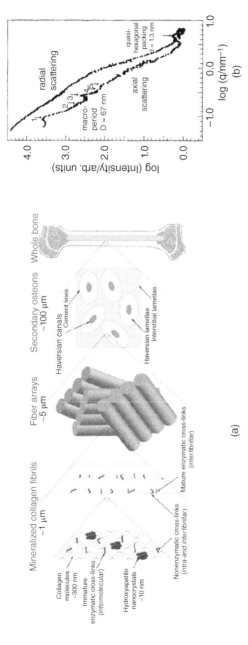

**Figure 4.28** (a) Simplified schematic of some levels of hierarchical order in bone that contains mineralised collagen fibres. *Source*: From Ref. [142]), (b) radial and axial SAXS patterns from bone, ulna from an adult mouse. Reflections 1–5 from the collagen long period $d = 67$ nm are indicated, along with a peak at high $q$ corresponding to $d = 1.3$ nm, related to the inter-planar spacing within the hexagonal packing of the collagen chains. *Source*: From Ref. [143].

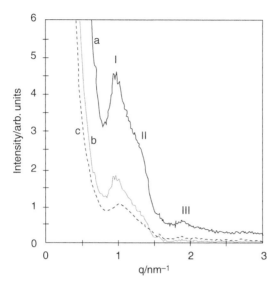

**Figure 4.29** SAXS intensity profiles measured for human stratum corneum. The peaks from the lipid bilayer structure are labelled I and III (first and third orders) with a shoulder peak II. Data from a 40% hydrated sample is shown with a solid line (labelled a), from a 20% hydrated sample with a dashed line (labelled b) and from the dry stratum corneum with a dotted line (labelled c). *Source*: From Bouwstra et al. [154]. © 1991, Elsevier.

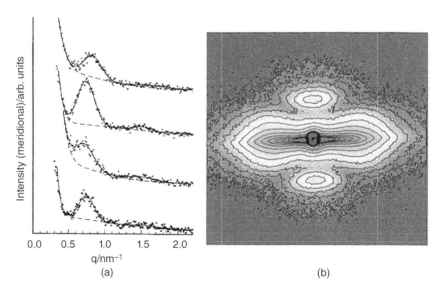

**Figure 4.30** (a) SAXS intensity profiles (along meridional direction) from spider dragline silk samples. SAXS profiles showing Bragg reflections from the semicrystalline lamellar structure under different sample conditions of temperature, hydration, and tensile deformation, (b) SAXS pattern from an aligned hydrated sample, after release of extension. Note the aligned meridional Bragg peaks. *Source*: From Ref. [147].

second order) are due to the lamellar structure, and the position can be mapped depending on the sample conditions. Aligned silk fibres show oriented SAXS patterns, an example being shown in Figure 4.30b for a hydrated sample that was stretched, SAXS being measured after release of the tensile force [157]. WAXS provides information on the packing of the chains (β-strands) in the crystallites and their alignment [158, 159].

Microfocus x-ray diffraction (microfocus WAXS) has been performed during in situ extension of a fine spider silk strand (7 μm beam focus), in order to probe the structure of the crystallites as a function of strain (and hydration level) [160]. Scanning WAXS has also been performed to map the texture within different types of spider silk threads [159].

Cellulose is the most abundant biomaterial. Its structure, which in trees and other plants comprises bundles of fibrils, has been investigated by SAXS and SAXS/WAXS. The structure of cellulose comprises arrays of microfibrils, the packing of which depends on the type of plant. Acid treatment of cellulose produces cellulose microcrystals that have a rod-shape and are sometimes termed *whiskers* that form colloidal suspensions, the SAXS patterns from which have been investigated [161].

Comparing acid (sulphite) treated wood and untreated wood, Figure 4.31 shows examples of SAXS patterns measured from different types of wood

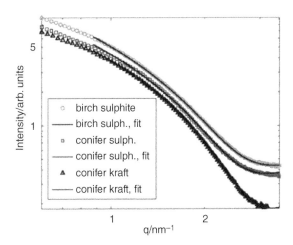

**Figure 4.31** SAXS data comparing of wood samples (hydrated specimens in sealed capillaries) comparing different types of wood, and also comparing acid (sulphite)-treated and untreated (Kraft cooked) samples of conifer wood. Open symbols are measured data and lines are fitted form factors – circular cylinder form factors for the woods labelled conifer and elliptical cylinder for the birch sample. *Source*: From Leppanen et al. [162]. © 2009, Springer Nature.

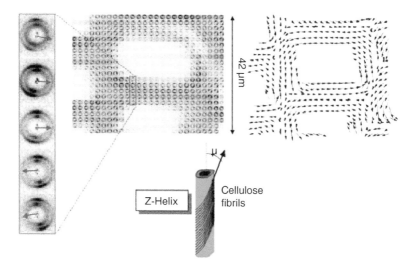

**Figure 4.32** Scanning microfocus WAXS patterns from a 2 mm-thick sample of spruce wood mounted in resin. The orientation of the 110 and 020 reflections (shown in enlarged WAXS patterns on the left) is mapped to show the variation in cellulose fibril orientation around wood cells, shown by the plot with arrows on the right-hand side. *Sources*: From Fratzl and Weinkamer [153]. © 2007, Elsevier; Based on data in Lichtenegger et al. [163].

pulp that can be described using circular or elliptical fibre form factors [162]. In the same report, the crystallinity and the dimensions of the cellulose microcrystals were obtained from WAXS [162]. As well as wood cellulose, this report also presents SAXS and WAXS data for flax and cotton cellulose.

Scanning microfocus WAXS (Section 3.11) has been used to investigate the variation in orientation of cellulose within a sample. Figure 4.32 shows an example of such measurements for a sample of Norwegian spruce wood [163]; the mapping of cellulose microfibril alignment around the wood cells is clearly defined.

## 4.16   FAST TIME-RESOLVED SAXS

Synchrotron SAXS instruments now permit sub-ms time-resolved measurements on many systems, for example soft materials that undergo self-assembly or the assembly of biomolecule complexes. These transitions may be induced using stopped flow methods, temperature jumps, or other perturbations to the system discussed in Section 3.12. There are now large numbers of fast time-resolved kinetic studies on many systems as reviewed elsewhere [164–166]. Here, a few recent examples are provided. The first is from a stopped flow SAXS study (setup shown in Figure 4.33a) of a charge-induced switch in conformation of a single-stranded DNA origami structure [167]. The X-shaped DNA origami construct undergoes a transition to a linear closed conformation at high divalent salt concentration [$MgCl_2$] (Figure 4.33b) as shown by time-resolved SAXS with an exposure time $t_e$ of 10 ms (Figure 4.33c). The SAXS data was analysed using the initial and final SAXS patterns as reference points to determine the kinetics of the formation of the fraction of closed molecules (Figure 5.27d), which follows first-order kinetics. Figure 4.33c includes a plot of $qI(q)$ versus $q$ (a so-called Holzer plot) consistent with Eq. (1.93) for a rod-like structure, which shows the development of a peak associated with the linear closed conformation.

The second example shows the use of microfluidics and microfocus SAXS along with sub-ms time resolution, pushing the limits of kinetic measurements currently achievable at synchrotron BioSAXS beamlines (the experiments were performed at the BioCAT beamline of the Advanced Photon Source). The refolding of the protein heart cytochrome $c$ was studied by dilution of the denaturing agent guanidinium hydrochloride (GdnHCl) [168]. The measurements were performed using a continuous flow microfluidic device (microfluidic cells are discussed in more detail in Section 3.12.6). The beam size at the sample was ~20 μm × ~5 μm full width at half maximum (FWHM) horizontal and vertical (microfocus SAXS and WAXS is discussed in Section 3.11). Figure 4.34 shows SAXS data measured after 100 μs (Figure 4.34a) along with Guinier plot analysis (Figure 4.34b), and a series of intensity profiles as the refolding proceeds (Figure 4.34c). The Kratky plot analysis presented in Figure 4.28d clearly shows the difference in conformation between unfolded (random coil) and folded states consistent with the analysis method discussed in Section 2.14.

The final example is a rare example of ps-time-resolved SAXS/WAXS using a free-electron laser (FEL) to study the photolysis of myoglobin in solution [169]. Figure 4.35 shows the SAXS/WAXS data with clear changes in both data sets even after 1 ps of exposure to 30 fs pulses from a 9 keV FEL x-ray beam. This leads to photolysis of the bond between the myoglobin and the CO ligand, and to disruption of the haem binding site and the surrounding protein chains, a so-called protein-quake, as illustrated schematically in Figure 4.35c. These state-of-the-art measurements highlight the potential for ultrafast SAXS measurements using free electron laser x-ray sources to study photolysis and other degradation mechanisms of molecules in very high intensity beams.

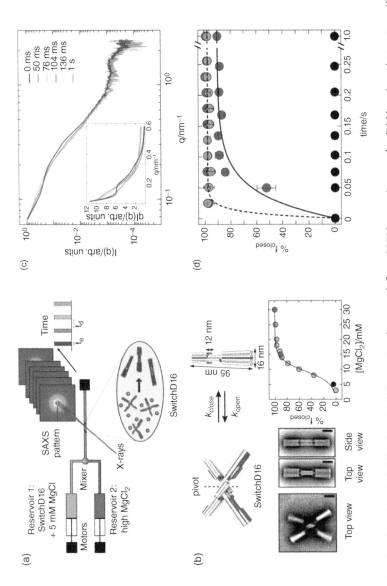

**Figure 4.33** DNA origami switch kinetics. (a) Schematic of stopped flow SAXS experiment, the DNA origami structure (SwitchD16, shown in part b) is mixed with a concentrated MgCl$_2$ solution, the SAXS data are measured with an exposure time $t_e$ (delay time $t_d$), (b) schematic of conformational switch of the DNA origami construct along with negative stain TEM images of the construct in the presence of 5 mM (left) and 25 mM (right) MgCl$_2$ (scale bar 20 nm) and fraction of closed structures as a function of [MgCl$_2$] determined from the SAXS analysis (coloured points relate to the analysis shown in part d) via a two-state model, (c) SAXS data at the times indicated, inset: Holzer plot of the SAXS data, (d) fraction of closed structures as a function of time obtained from a two-state analysis of the SAXS data at four [MgCl$_2$] concentrations, 5 mM (black), 15 mM (blue), 25 mM (orange) and 35 mM (red). *Source:* From Bruetzel et al. [167]. © 2018, American Chemical Society.

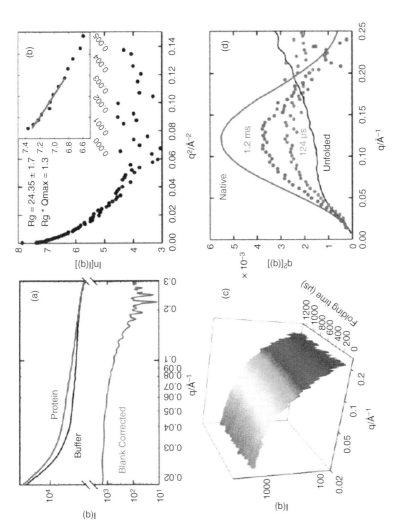

**Figure 4.34** Time-resolved microfocus and microfluidic SAXS study of the refolding of heart cytochrome *c* was studied by dilution of GdnHCl from the random coil state in 4.5 mM GdnHCl. (a) SAXS data from sample and buffer and subtracted data measured after 100 μs, (b) Guinier plot of SAXS data, providing radius of gyration indicated (inset: expansion of linear region at low *q*), (c) time-dependent SAXS profiles reaching final dilution conditions of 0.45 mM GdnHCl, (d) Kratky plots of data at two selected time points indicated along with equilibrium condition measurements for folded (red) and unfolded (black) protein. *Source:* From Graceffa et al [168]. © 2013, International Union of Crystallography.

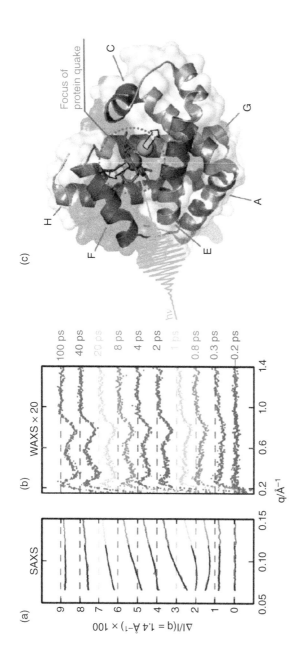

**Figure 4.35** Example of ultrafast time-resolved SAXS used in a study of the photolysis of a myoglobin-CO complex. Changes at the time points indicated in (a) SAXS and (b) WAXS intensity profiles for the complex in solution, represented as the ratio of the difference in intensity from initial value (at $t = -0.2$ ps before photolysis), with respect to measured intensity profile, the data being normalized using the intensity at $q = 1.4\,Å^{-1}$. The black lines in part (a) represent Guinier fits of the SAXS data. Data are offset vertically for clarity. (c) Model for the disruption of the protein structure (so-called 'protein quake', characterized by a 4 fs collective vibrational mode) following photolysis after exposure to the pulsed x-ray beam (green). The CO is represented by green and red spheres, the haem molecule by a red framework structure which is covalently bound to protein helix F and to parts of helices A, C, E, G, and H. *Source:* From Levantino et al. [169]. Licensed under CC BY 4.0.

# REFERENCES

1. Center for X-ray Optics and Advanced Light Source, Lawrence Berkeley National Laboratory.(2020). *X-ray Data Booklet*. http://xdb.lbl.gov.
2. Svergun, D.I., Koch, M.H.J., Timmins, P.A., and May, R.P. (2013). *Small Angle X-Ray and Neutron Scattering from Solutions of Biological Macromolecules*. Oxford: Oxford University Press.
3. Narayanan, T. (2014). Small-Angle Scattering. In: *Structure from Diffraction Methods* (eds. D.W. Bruce, D. O'Hare and R.I. Walton). Wiley Online Library: Wiley.
4. Roe, R.-J. (2000). *Methods of X-Ray and Neutron Scattering in Polymer Science*. New York: Oxford University Press.
5. Thompson, A.C., Kirz, J., Attwood, D.T. et al. (2009). *X-Ray Data Booklet*. Lawrence Berkeley National Laboratory.
6. Als-Nielsen, J. and McMorrow, D. (2001). *Elements of Modern X-Ray Physics*. Chichester: Wiley.
7. Lindner, P. (2002). Scattering experiments. In: *Neutrons, X-Rays and Light Scattering Methods Applied to Soft Condensed Matter* (eds. P. Lindner and T. Zemb). Amsterdam: Elsevier.
8. Prince, H.M. (ed.) (2004). *International Tables for Crystallography, Volume C*. Dordrecht: Kluwer.
9. Henke, B.L., Gullikson, E.M., and Davis, J.C. (1993). X-ray interactions - photoabsorption, scattering, transmission, and reflection at E=50-30,000 eV, Z=1-92. *Atomic Data and Nuclear Data Tables* 54: 181–342.
10. Sztucki, M., Di Cola, E., and Narayanan, T. (2012). Anomalous small-angle X-ray scattering from charged soft matter. *European Physical Journal Special Topics* 208: 319–331.
11. Cao, Y., Alaasar, M., Nallapaneni, A. et al. (2020). Molecular packing in double gyroid cubic phases revealed via resonant soft X-ray scattering. *Physical Review Letters* 125: 1–6.
12. Glatter, O. (2018). *Scattering Methods and Their Application in Colloid and Interface Science*, Chap.6. Amsterdam: Elsevier.
13. Doniach, S. (2001). Changes in biomolecular conformation seen by small angle X-ray scattering. *Chemical Reviews* 101: 1763–1778.
14. Svergun, D.I. and Koch, M.H.J. (2003). Small-angle scattering studies of biological macromolecules in solution. *Reports on Progress in Physics* 66: 1735–1782.
15. Putnam, C.D., Hammel, M., Hura, G.L., and Tainer, J.A. (2007). X-ray solution scattering (SAXS) combined with crystallography and computation: defining accurate macromolecular structures, conformations and assemblies in solution. *Quarterly Reviews of Biophysics* 40: 191–285.
16. Doniach, S. and Lipfert, J. (2007). Small and wide angle X-ray scattering from biological macromolecules and their complexes in solution. In: *Comprehensive Biophysics* (ed. E.H. Egelman), 376–397. New York: Academic Press.
17. Lipfert, J. and Doniach, S. (2007). Small-angle X-ray scattering from RNA, proteins, and protein complexes. *Annual Review of Biophysics and Biomolecular Structure* 36: 307–327.
18. Mertens, H.D.T. and Svergun, D.I. (2010). Structural characterization of proteins and complexes using small-angle X-ray solution scattering. *Journal of Structural Biology* 172: 128–141.
19. Jacques, D.A. and Trewhella, J. (2010). Small-angle scattering for structural biology-expanding the frontier while avoiding the pitfalls. *Protein Science* 19: 642–657.
20. Blanchet, C.E. and Svergun, D.I. (2013). Small-angle X-ray scattering on biological macromolecules and nanocomposites in solution. In: *Annual Review of Physical Chemistry*, vol. 64 (eds. M.A. Johnson and T.J. Martinez), 37–54. Palo Alto: Annual Reviews.

21. Chaudhuri, B., Muñoz, I.G., Qian, S., and Urban, V.S. (eds.) (2017). *Biological Small Angle Scattering: Techniques, Strategies and Tips*. Singapore: Springer.

22. Lattman, E.E., Grant, T.D., and Snell, E.H. (2018). *Biological Small-Angle Scattering: Theory and Practice*. Oxford: Oxford University Press.

23. Delagenière, S., Brenchereau, P., Ludovic Launer, L. et al. (2011). ISPyB: an information management system for synchrotron macromolecular crystallography. *Bioinformatics* 27: 3186–3192.

24. Svergun, D.I. (1992). Determination of the regularization parameter in indirect-transform methods using perceptual criteria. *Journal of Applied Crystallography* 25: 495–503.

25. Glatter, O. (1977). A new method for the evaluation of small-angle scattering data. *Journal of Applied Crystallography* 10: 415–421.

26. Brunner-Popela, J. and Glatter, O. (1997). Small-angle scattering of interacting particles .1. Basic principles of a global evaluation technique. *Journal of Applied Crystallography* 30: 431–442.

27. Bergmann, A., Fritz, G., and Glatter, O. (2000). Solving the generalized indirect Fourier transformation (GIFT) by Boltzmann simplex simulated annealing (BSSA). *Journal of Applied Crystallography* 33: 1212–1216.

28. Fritz, G., Bergmann, A., and Glatter, O. (2000). Evaluation of small-angle scattering data of charged particles using the generalized indirect Fourier transformation technique. *Journal of Chemical Physics* 113: 9733–9740.

29. Glatter, O. (2018). *Scattering Methods and Their Application in Colloid and Interface Science*, Chap.3. Amsterdam: Elsevier.

30. Fraser, R.D.B., MacRae, T.P., and Suzuki, E. (1978). An improved method for calculating the contribution of solvent to the X-ray diffraction pattern of biological molecules. *Journal of Applied Crystallography* 11: 693–694.

31. Lattman, E.E. (1989). Rapid calculation of the solution scattering profile from a macromolecule of known structure. *Proteins: Structure, Function, and Bioinformatics* 5: 149–155.

32. Svergun, D., Barberato, C., and Koch, M.H.J. (1995). CRYSOL - a program to evaluate x-ray solution scattering of biological macromolecules from atomic coordinates. *Journal of Applied Crystallography* 28: 768–773.

33. Schneidman-Duhovny, D., Hammel, M., Tainer, J.A., and Sali, A. (2013). Accurate SAXS profile computation and its assessment by contrast variation experiments. *Biophysical Journal* 105: 962–974.

34. Lee, B. and Richards, F.M. (1971). The interpretation of protein structures: estimation of static accessibility. *Journal of Molecular Biology* 55: 379–400.

35. Shrake, A. and Rupley, J.A. (1973). Environment and exposure to solvent of protein atoms. Lysozyme and insulin. *Journal of Molecular Biology* 79: 351–364.

36. Connolly, M.L. (1983). Analytical molecular surface calculation. *Journal of Applied Crystallography* 16: 548–558.

37. Connolly, M.L. (1983). Solvent-accessible surfaces of proteins and nucleic acids. *Science* 221: 709–713.

38. Connolly, M.L. (1985). Molecular surface triangulation. *Journal of Applied Crystallography* 18: 499–505.

39. Poitevin, F., Orland, H., Doniach, S. et al. AquaSAXS: a web server for computation and fitting of SAXS profiles with non-uniformly hydrated atomic models. *Nucleic Acids Research* 39: W184–W189.

40. Grishaev, A., Guo, L.A., Irving, T., and Bax, A. (2010). Improved fitting of solution X-ray scattering data to macromolecular structures and structural ensembles by explicit water modeling. *Journal of the American Chemical Society* 132: 15484–15486.

41. Liu, H.G., Hexemer, A., and Zwart, P.H. (2012). The Small Angle Scattering ToolBox (SASTBX): an open-source software for biomolecular small-angle scattering. *Journal of Applied Crystallography* 45: 587–593.

42. Rambo, R.P. and Tainer, J.A. (2013). Super-resolution in solution X-ray scattering and its applications to structural systems biology. In: *Annual Review of Biophysics*, vol. 42 (ed. K.A. Dill), 415–441. Palo Alto: Annual Reviews.

43. Hammel, M. (2012). Validation of macromolecular flexibility in solution by small-angle X-ray scattering (SAXS). *European Biophysics Journal with Biophysics Letters* 41: 789–799.

44. Müller, J.J., Hansen, S., and Pürschel, H.V. (1996). The use of small-angle scattering and the maximum-entropy method for shape-model determination from distance-distribution functions. *Journal of Applied Crystallography* 29: 547–554.

45. Stuhrmann, H.B. (1970). *Acta Crystallographica* A26: 297–306.

46. Stuhrmann, H.B. (1970). Ein neues Verfahren zur Bestimmung der Oberflächenform und der inneren Struktur von gelösten globulären Proteinen aus Röntgenkleinwinkelmessungen. *Zeitschrift für Physikalische Chemie-Frankfurt* 72: 177. ff.

47. Konarev, P.V., Petoukhov, M.V., and Svergun, D.I. (2001). MASSHA - a graphic system for rigid body modelling of macromolecular complexes against solution scattering data. *Journal of Applied Crystallography* 34: 527–532.

48. Svergun, D.I. (1999). Restoring low resolution structure of biological macromolecules from solution scattering using simulated annealing. *Biophysical Journal* 76: 2879–2886.

49. Franke, D., Petoukhov, M.V., Konarev, P.V. et al. (2017). ATSAS 2.8: a comprehensive data analysis suite for small-angle scattering from macromolecular solutions. *Journal of Applied Crystallography* 50: 1212–1225.

50. Petoukhov, M.V. and Svergun, D.I. (2005). Global rigid body modeling of macromolecular complexes against small-angle scattering data. *Biophysical Journal* 89: 1237–1250.

51. Svergun, D.I., Petoukhov, M.V., and Koch, M.H.J. (2001). Determination of domain structure of proteins from x-ray solution scattering. *Biophysical Journal* 80: 2946–2953.

52. Perkins, S.J., Okemefuna, A.I., Fernando, A.N. et al. (2008). X-ray and neutron scattering data and their constrained molecular modeling. In: *Biophysical Tools for Biologists: Vol 1 In Vitro Techniques. Methods in Cell Biology*, vol. 84 (eds. J.J. Correia and H.W. Detrich), 375–423. San Diego: Elsevier Academic Press Inc.

53. Schneidman-Duhovny, D., Hammel, M., and Sali, A. (2011). Macromolecular docking restrained by a small angle X-ray scattering profile. *Journal of Structural Biology* 173: 461–471.

54. Schneidman-Duhovny, D., Hammel, M., Tainer, J.A., and Sali, A. (2016). FoXS, FoXS-Dock and MultiFoXS: single-state and multi-state structural modeling of proteins and their complexes based on SAXS profiles. *Nucleic Acids Research* 44: W424–W429.

55. Pelikan, M., Hura, G.L., and Hammel, M. (2009). Structure and flexibility within proteins as identified through small angle X-ray scattering. *General Physiology and Biophysics* 28: 174–189.

56. Förster, F., Webb, B., Krukenberg, K.A. et al. (2008). Integration of small-angle X-ray scattering data into structural modeling of proteins and their assemblies. *Journal of Molecular Biology* 382: 1089–1106.

57. Rozycki, B., Kim, Y.C., and Hummer, G. (2011). SAXS ensemble refinement of ESCRT-III CHMP3 conformational transitions. *Structure* 19: 109–116.

58. Schwieters, C.D., Suh, J.Y., Grishaev, A. et al. (2010). Solution structure of the 128 kDa enzyme I dimer from Escherichia coli and its 146 kDa complex with HPr using residual dipolar couplings and small- and wide-angle X-ray scattering. *Journal of the American Chemical Society* 132: 13026–13045.

59. Boehm, M.K., Woof, J.M., Kerr, M.A., and Perkins, S.J. (1999). The fab and fc fragments of IgA1 exhibit a different arrangement from that in IgG: a study by X-ray and neutron solution scattering and homology modelling. *Journal of Molecular Biology* 286: 1421–1447.

60. Hammel, M., Fierober, H.P., Czjzek, M. et al. (2005). Structural basis of cellulosome efficiency explored by small angle X-ray scattering. *Journal of Biological Chemistry* 280: 38562–38568.

61. Yang, S.C., Blachowicz, L., Makowski, L., and Roux, B. (2010). Multidomain assembled states of Hck tyrosine kinase in solution. *Proceedings of the National Academy of Sciences of the United States of America* 107: 15757–15762.

62. Petoukhov, M.V., Franke, D., Shkumatov, A.V. et al. (2012). New developments in the ATSAS program package for small-angle scattering data analysis. *Journal of Applied Crystallography* 45: 342–350.

63. Konarev, P.V., Volkov, V.V., Sokolova, A.V. et al. (2003). PRIMUS: a windows PC-based system for small-angle scattering data analysis. *Journal of Applied Crystallography* 36: 1277–1282.

64. Vestergaard, B., Groenning, M., Roessle, M. et al. (2007). A helical structural nucleus is the primary elongating unit of insulin amyloid fibrils. *PLoS Biology* 5: 1089–1097.

65. Volkov, V.V. and Svergun, D.I. (2003). Uniqueness of ab initio shape determination in small-angle scattering. *Journal of Applied Crystallography* 36: 860–864.

66. Giehm, L., Svergun, D.I., Otzen, D.E., and Vestergaard, B. (2011). Low-resolution structure of a vesicle disrupting alpha-synuclein oligomer that accumulates during fibrillation. *Proceedings of the National Academy of Sciences of the United States of America* 108: 3246–3251.

67. Asor, R., Schlicksup, J., Zhao, Z. et al. (2020). Rapidly forming early intermediate structures dictate the pathway of capsid assembly. *Journal of the American Chemical Society* 142: 7868–7882.

68. Zhang, F.J., Skoda, M.W.A., Jacobs, R.M.J. et al. (2007). Protein interactions studied by SAXS: effect of ionic strength and protein concentration for BSA in aqueous solutions. *Journal of Physical Chemistry. B* 111: 251–259.

69. Stell, G. (1963). The Percus-Yevick Equation for the radial distribution function of a fluid. *Physica* 29: 517–534.

70. Hansen, J.P. and Macdonald, I.R. (2013). *Theory of Simple Liquids*, 4th edition. London: Academic Press.

71. Stradner, A., Sedgwick, H., Cardinaux, F. et al. (2004). Equilibrium cluster formation in concentrated protein solutions and colloids. *Nature* 432: 492–495.

72. Liu, Y. and Xi, Y.Y. (2019). Colloidal systems with a short-range attraction and long-range repulsion: phase diagrams, structures, and dynamics. *Current Opinion in Colloid and Interface Science* 39: 123–136.

73. van Heijkamp, L.F., de Schepper, I.M., Strobl, M. et al. (2010). Milk gelation studied with small angle neutron scattering techniques and Monte Carlo simulations. *Journal of Physical Chemistry. A* 114: 2412–2426.

74. Pignon, F., Belina, G., Narayanan, T. et al. (2004). Structure and rheological behavior of casein micelle suspensions during ultrafiltration process. *Journal of Chemical Physics* 121: 8138–8146.

75. Marchin, S., Putaux, J.L., Pignon, F., and Leonil, J. (2007). Effects of the environmental factors on the casein micelle structure studied by cryo transmission electron microscopy and small-angle x-ray scattering/ultrasmall-angle x-ray scattering. *Journal of Chemical Physics* 126: 10.

76. Tromp, R.H. and Bouwman, W.G. (2007). A novel application of neutron scattering on dairy products. *Food Hydrocolloids* 21: 154–158.

77. Hamley, I.W. (2007). *Introduction to Soft Matter*, Revised edition. Chichester: Wiley.

78. Jones, R.A.L. (2002). *Soft Condensed Matter*. Oxford: Oxford University Press.

79. Doi, M. (2013). *Soft Matter Physics*. Oxford: Oxford University Press.

80. Terentjev, E.M. and Weitz, D.A. (eds.) (2018). *The Oxford Handbook of Soft Condensed Matter*. Oxford University Press.

81. Ballauff, M. (2001). SAXS and SANS studies of polymer colloids. *Current Opinion in Colloid and Interface Science* 6: 132–139.

82. Hamley, I.W. and Castelletto, V. (2004). Small-angle scattering of block copolymers in the melt, solution and crystal states. *Progress in Polymer Science* 29: 909–948.

83. Stribeck, N. (2007). *X-Ray Scattering of Soft Matter*. Berlin: Springer-Verlag.

84. Narayanan, T. (2009). High brilliance small-angle x-ray scattering applied to soft matter. *Current Opinion in Colloid & Interface Science* 14: 409–415.

85. de Jeu, W.H. (2016). *Basic X-Ray Scattering for Soft Matter*. Oxford: Oxford University Press.

86. Narayanan, T., Wacklin, H., Konovalov, O., and Lund, R. (2017). Recent applications of synchrotron radiation and neutrons in the study of soft matter. *Crystallography Reviews* 23: 160–226.

87. Ryan, A.J., Hamley, I.W., Bras, W., and Bates, F.S. (1995). Structure development in semicrystalline diblock copolymers crystallizing from the ordered melt. *Macromolecules* 28: 3860–3868.

88. Balta-Calleja, F.J. and Vonk, C.G. (1989). *X-Ray Scattering of Synthetic Polymers*. London: Elsevier.

89. Strobl, G.R. (1996). *The Physics of Polymers. Concepts for Understanding their Structures and Behavior*. Berlin: Springer.

90. Strobl, G.R. and Schneider, M. (1980). Direct evaluation of the electron-density correlation-function of partially crystalline polymers. *Journal of Polymer Science Part B-Polymer Physics* 18: 1343–1359.

91. Warren, B.E. (1969). *X-Ray Diffraction*. Reading, MA: Addison-Wesley.

92. Worthington, C.R., King, G.I., and McIntosh, T.J. (1973). Direct structure determination of multilayered membrane-type systems which contain fluid layers. *Biophysical Journal* 13: 480–494.

93. McIntosh, T.J., Magid, A.D., and Simon, S.A. (1989). Range of the solvation pressure between lipid-membranes - dependence on the packing density of solvent molecules. *Biochemistry* 28: 7904–7912.

94. Yu, Z.W. and Quinn, P.J. (1995). Phase-stability of phosphatidylcholines in dimethylsulfoxide solutions. *Biophysical Journal* 69: 1456–1463.

95. Nagle, J.F. and Tristram-Nagle, S. (2000). Structure of lipid bilayers. *Biochimica et Biophysica Acta* 1469: 159–195.

96. Harper, P.E., Mannock, D.A., Lewis, R. et al. (2001). X-ray diffraction structures of some phosphatidylethanolamine lamellar and inverted hexagonal phases. *Biophysical Journal* 81: 2693–2706.

97. Tyler, A.I.I., Law, R.V., and Seddon, J.M. (2015). X-ray diffraction of lipid model membranes. In: *Methods in Membrane Lipids* (ed. D.M. Owen). New York: Springer.

98. Torbet, J. and Wilkins, M.H.F. (1976). X-ray-diffraction studies of lecithin bilayers. *Journal of Theoretical Biology* 62: 447–458.

99. Sato, K., Arishima, T., Wang, Z.H. et al. (1989). Polymorphism of POP and SOS. 1. Occurrence and polymorphic transformation. *Journal of the American Oil Chemists' Society* 66: 664–674.

100. Ueno, S., Minato, A., Seto, H. et al. (1997). Synchrotron radiation x-ray diffraction study of liquid crystal formation and polymorphic crystallization of SOS (*sn*-1,-3-distearoyl-2-oleoyl glycerol). *The Journal of Physical Chemistry. B* 101: 6847–6854.

101. van Malssen, K., van Langevelde, A., Peschar, R., and Schenk, H. (1999). Phase behavior and extended phase scheme of static cocoa butter investigated with real-time X-ray powder diffraction. *Journal of the American Oil Chemists' Society* 76: 669–676.

102. Mykhaylyk, O.O. and Hamley, I.W. (2004). The packing of triacylglycerols from SAXS measurements: application to the structure of 1,3-distearoyl-2-oleoyl-*sn*-glycerol crystal phases. *Journal of Physical Chemistry. B* 108: 8069–8083.

103. Hirai, M., Iwase, H., Hayakawa, T. et al. (2003). Determination of asymmetric structure of ganglioside-DPPC mixed vesicle using SANS, SAXS, and DLS. *Biophysical Journal* 85: 1600–1610.

104. Brzustowicz, M.R. and Brunger, A.T. (2005). X-ray scattering from unilamellar lipid vesicles. *Journal of Applied Crystallography* 38: 126–131.

105. Kucerka, N., Pencer, J., Sachs, J.N. et al. (2007). Curvature effect on the structure of phospholipid bilayers. *Langmuir* 23: 1292–1299.

106. Eicher, B., Heberle, F.A., Marquardt, D. et al. (2017). Joint small-angle x-ray and neutron scattering data analysis of asymmetric lipid vesicles. *Journal of Applied Crystallography* 50: 419–429.

107. Komorowski, K., Salditt, A., Xu, Y.H. et al. (2018). Vesicle adhesion and fusion studied by small-angle x-ray scattering. *Biophysical Journal* 114: 1908–1920.

108. Hamley, I.W., Castelletto, V., Dehsorkhi, A. et al. (2018). The conformation and aggregation of proline-rich surfactant-like peptides. *Journal of Physical Chemistry. B* 122: 1826–1835.

109. Castelletto, V., Barnes, R.H., Karatzas, K.A. et al. (2018). Arginine-containing surfactant-like peptides: interaction with lipid membranes and antimicrobial activity. *Biomacromolecules* 19: 2782–7294.

110. Pabst, G., Rappolt, M., Amenitsch, H., and Laggner, P. (2000). Structural information from multilamellar liposomes at full hydration: full q-range fitting with high quality x-ray data. *Physical Review E* 62: 4000–4009.

111. Castelletto, V., Kaur, A., Kowalczyk, R.M. et al. (2017). Supramolecular hydrogel formation in a series of self-assembling lipopeptides with varying lipid chain length. *Biomacromolecules* 18: 2013–2023.

112. Hamley, I.W., Dehsorkhi, A., Castelletto, V. et al. (2013). Reversible helical ribbon unwinding transition of a self-assembling peptide amphiphile. *Soft Matter* 9: 9290–9293.

113. Hamley, I.W., Dehsorkhi, A., Jauregi, P. et al. (2013). Self-assembly of three bacterially-derived bioactive lipopeptides. *Soft Matter* 9: 9572–9578.

114. Miravet, J.F., Escuder, B., Segarra-Maset, M.D. et al. (2013). Self-assembly of a peptide amphiphile: transition from nanotape fibrils to micelles. *Soft Matter* 9: 3558–3564.

115. Dehsorkhi, A., Hamley, I.W., Seitsonen, J., and Ruokolainen, J. (2013). Tuning self-assembled nanostructures through enzymatic degradation of a peptide amphiphile. *Langmuir* 29: 6665–6672.

116. Hamley, I.W., Kirkham, S., Dehsorkhi, A. et al. (2014). Toll-like receptor agonist lipopeptides self-assemble into distinct nanostructures. *Chemical Communications* 50: 15948–15951.

117. Dehsorkhi, A., Gouveia, R.J., Smith, A.M. et al. (2015). Self-assembly of a dual functional bioactive peptide amphiphile incorporating both matrix metalloprotease substrate and cell adhesion motifs. *Soft Matter* 11: 3115–3124.

118. Hutchinson, J.A., Burholt, S., Hamley, I.W. et al. (2018). The effect of Lipidation on the self-assembly of the gut derived peptide hormone PYY3-36. *Bioconjugate Chemistry* 29: 2296–2308.

119. Castelletto, V., Hamley, I.W., Seitsonen, J. et al. (2018). Conformation and aggregation of selectively PEGylated and lipidated gastric peptide hormone human $PYY_{3-36}$. *Biomacromolecules* 19: 4320–4332.

120. Pedersen, J.S. (1997). Analysis of small-angle scattering data from colloids and polymer solutions: modeling and least-squares fitting. *Advances in Colloid and Interface Science* 70: 171–210.

121. Hamley, I.W., Krysmann, M.J., Castelletto, V. et al. (2008). Nematic and columnar ordering of a PEG-peptide conjugate in aqueous solution. *Chemistry - A European Journal* 14: 11369–11374.

122. Hamley, I.W., Krysmann, M.J., Castelletto, V., and Noirez, L. (2008). Multiple lyotropic polymorphism of a PEG-peptide diblock copolymer in aqueous solution. *Advanced Materials* 20: 4394–4397.

123. Castelletto, V., Cheng, G., Stain, C. et al. (2012). Self-assembly of a peptide amphiphile containing L-carnosine and its mixtures with a multilamellar vesicle forming lipid. *Langmuir* 28: 11599–11608.

124. Hamley, I.W., Dehsorkhi, A., and Castelletto, V. (2013). Self-assembled arginine-coated peptide nanosheets in water. *Chemical Communications* 49: 1850–1852.

125. Castelletto, V., Gouveia, R.J., Connon, C.J., and Hamley, I.W. (2013). New RGD- peptide amphiphile mixtures containing a negatively charged diluent. *Faraday Discussions* 166: 381–397.

126. Caillé, A. (1972). X-ray scattering by smectic-a crystals. *Comptes Rendus Hebdomadaires des Seances de l'Academie des Sciences Serie B* 274: 891–892.

127. Lu, K., Jacob, J., Thiyagarajan, P. et al. (2003). Exploiting amyloid fibril lamination for nanotube self assembly. *Journal of the American Chemical Society* 125: 6391–6393.

128. Hamley, I.W., Burholt, S., Hutchinson, J. et al. (2017). Shear alignment of bola-amphiphilic arginine-coated peptide nanotubes. *Biomacromolecules* 18: 141–149.

129. Guilbaud, J.B. and Saiani, A. (2011). Using small angle scattering (SAS) to structurally characterise peptide and protein self-assembled materials. *Chemical Society Reviews* 40: 1200–1210.

130. Li, L., Harnau, L., Rosenfeldt, S., and Ballauff, M. (2005). Effective interaction of charged platelets in aqueous solution: investigations of colloid laponite suspensions by static light scattering and small-angle x-ray scattering. *Physical Review E* 72: 10.

131. Pontoni, D., Finet, S., Narayanan, T., and Rennie, A.R. (2003). Interactions and kinetic arrest in an adhesive hard-sphere colloidal system. *Journal of Chemical Physics* 119: 6157–6165.

132. Petukhov, A.V., van der Beek, D., Dullens, R.P.A. et al. (2005). Observation of a hexatic columnar liquid crystal of polydisperse colloidal disks. *Physical Review Letters* 95: 4.

133. Petukhov, A.V., Thijssen, J.H.J., 't Hart, D.C. et al. (2006). Microradian x-ray diffraction in colloidal photonic crystals. *Journal of Applied Crystallography* 39: 137–144.

134. Thijssen, J.H.J., Petukhov, A.V., t'Hart, D.C. et al. (2006). Characterization of photonic colloidal single crystals by microradian x-ray diffraction. *Advanced Materials* 18: 1662–1666.

135. Voet, D. and Voet, J.G. (1995). *Biochemistry*. New York: Wiley.

136. Shoulders, M.D. and Raines, R.T. (2009). Collagen structure and stability. *Annual Review of Biochemistry* 78: 929–958.

137. Fraser, R.D.B. and MacRae, T.P. (1981). Unit-cell and molecular connectivity in tendon collagen. *International Journal of Biological Macromolecules* 3: 193–200.

138. Orgel, J.P.R.O., Miller, A., Irving, T.C. et al. (2001). The in situ supermolecular structure of type I collagen. *Structure* 9: 1061–1069.

139. Orgel, J., Irving, T.C., Miller, A., and Wess, T.J. (2006). Microfibrillar structure of type I collagen in situ. *Proceedings of the National Academy of Sciences of the United States of America* 103: 9001–9005.

140. Nádaždy, P., Hagara, J. Jergel, M. Majková, E., Mikulík, P., Zápražný, Z., Korytár, D. and Šiffalovič, P. (2019). Exploiting the potential of beam-compressing channel-cut monochromators for laboratory high-resolution small-angle X-ray scattering experiments. *Journal of Applied Crystallography* 52: 498–506.

141. Hamley, I.W. (2020). *Introduction to Peptide Science*. Chichester: Wiley.

142. Zimmermann, E.A., Schaible, E., Bale, H. et al. (2011). Age-related changes in the plasticity and toughness of human cortical bone at multiple length scales. *Proceedings of the National Academy of Sciences of the United States of America* 108: 14416–14421.

143. Fratzl, P., Fratzl-Zelman, N., Klaushofer, K. et al. (1991). Nucleation and growth of mineral crystals in bone studied by small-angle x-ray-scattering. *Calcified Tissue International* 48: 407–413.

144. Gourrier, A., Wagermaier, W., Burghammer, M. et al. (2007). Scanning X-ray imaging with small-angle scattering contrast. *Journal of Applied Crystallography* 40: S78–S82.

145. Wagermaier, W., Gupta, H.S., Gourrier, A. et al. (2007). Scanning texture analysis of lamellar bone using microbeam synchrotron x-ray radiation. *Journal of Applied Crystallography* 40: 115–120.

146. Tanaka, T., Yagi, N., Ohta, T. et al. (2010). Evaluation of the distribution and orientation of Remineralized enamel crystallites in subsurface lesions by X-ray diffraction. *Caries Research* 44: 253–259.

147. Deyhle, H., White, S.N., Bunk, O. et al. (2014). Nanostructure of carious tooth enamel lesion. *Acta Biomaterialia* 10: 355–364.

148. Kinney, J.H., Pople, J.A., Marshall, G.W., and Marshall, S.J. (2001). Collagen orientation and crystallite size in human dentin: a small angle x-ray scattering study. *Calcified Tissue International* 69: 31–37.

149. Tesch, W., Eidelman, N., Roschger, P. et al. (2001). Graded microstructure and mechanical properties of human crown dentin. *Calcified Tissue International* 69: 147–157.

150. Saranathan, V., Osuji, C.O., Mochrie, S.G.J. et al. (2010). Structure, function, and self-assembly of single network gyroid (I4(1)32) photonic crystals in butterfly wing scales. *Proceedings of the National Academy of Sciences of the United States of America* 107: 11676–11681.

151. Vrieling, E.G., Beelen, T.P.M., van Santen, R.A., and Gieskes, W.W.C. (1999). Diatom silicon biomineralization as an inspirational source of new approaches to silica production. *Journal of Biotechnology* 70: 39–51.

152. Vrieling, E.G., Beelen, T.P.M., van Santen, R.A., and Gieskes, W.W.C. (2000). Nanoscale uniformity of pore architecture in diatomaceous silica: a combined small and wide angle x-ray scattering study. *Journal of Phycology* 36: 146–159.

153. Fratzl, P. and Weinkamer, R. (2007). Nature's hierarchical materials. *Progress in Materials Science* 52: 1263–1334.

154. Bouwstra, J.A., Devries, M.A., Gooris, G.S. et al. (1991). Thermodynamic and structural aspects of the skin barrier. *Journal of Controlled Release* 15: 209–220.

155. Bouwstra, J.A., Gooris, G.S., Vanderspek, J.A., and Bras, W. (1991). Structural investigations of human stratum-corneum by small-angle x-ray-scattering. *Journal of Investigative Dermatology* 97: 1005–1012.

156. Creighton, T.E. (1993). *Proteins. Structures and Molecular Properties*. New York: W.H. Freeman.

157. Yang, Z., Grubb, D.T., and Jelinski, L.W. (1997). Small-angle x-ray scattering of spider dragline silk. *Macromolecules* 30: 8254–8261.

158. Riekel, C. and Vollrath, F. (2001). Spider silk fibre extrusion: combined wide- and small-angle x-ray microdiffraction experiments. *International Journal of Biological Macromolecules* 29: 203–210.

159. Riekel, C. (2003). Applications of micro-SAXS/WAXS to study polymer fibers. *Nuclear Instruments & Methods in Physics Research Section B-Beam Interactions with Materials and Atoms* 199: 106–111.

160. Glisovic, A., Vehoff, T., Davies, R.J., and Salditt, T. (2008). Strain dependent structural changes of spider dragline silk. *Macromolecules* 41: 390–398.

161. Lima, M.M.D. and Borsali, R. (2004). Rodlike cellulose microcrystals: structure, properties, and applications. *Macromolecular Rapid Communications* 25: 771–787.

162. Leppanen, K., Andersson, S., Torkkeli, M. et al. (2009). Structure of cellulose and microcrystalline cellulose from various wood species, cotton and flax studied by x-ray scattering. *Cellulose* 16: 999–1015.

163. Lichtenegger, H., Muller, M., Paris, O. et al. (1999). Imaging of the helical arrangement of cellulose fibrils in wood by synchrotron x-ray microdiffraction. *Journal of Applied Crystallography* 32: 1127–1133.

164. Panine, P., Finet, S., Weiss, T.M., and Narayanan, T. (2006). Probing fast kinetics in complex fluids by combined rapid mixing and small-angle x-ray scattering. *Advances in Colloid and Interface Science* 127: 9–18.

165. Levantino, M., Yorke, B.A., Monteiro, D.C.F. et al. (2015). Using synchrotrons and XFELs for time-resolved x-ray crystallography and solution scattering experiments on biomolecules. *Current Opinion in Structural Biology* 35: 41–48.

166. Narayanan, T. and Konovalov, O. (2020). Synchrotron scattering methods for nanomaterials and soft matter research. *Materials* 13: 752.

167. Bruetzel, L.K., Walker, P.U., Gerling, T. et al. (2018). Time-resolved small-angle x-ray scattering reveals millisecond transitions of a DNA origami switch. *Nano Letters* 18: 2672–2676.
168. Graceffa, R., Nobrega, R.P., Barrea, R.A. et al. (2013). Sub-millisecond time-resolved SAXS using a continuous-flow mixer and x-ray microbeam. *Journal of Synchrotron Radiation* 20: 820–825.
169. Levantino, M., Schiro, G., Lemke, H.T. et al. (2015). Ultrafast myoglobin structural dynamics observed with an x-ray free-electron laser. *Nature Communications* 6: 6.

# 5

# Applications and Specifics of SANS

## 5.1 INTRODUCTION

In this chapter, the essential features of small-angle neutron scattering (SANS) are introduced, including SANS-specific details on scattering lengths and scattering cross-sections, the production of neutrons and SANS-specific measurement considerations. Then examples of SANS measurements on different systems are presented, highlighting the advantages and capabilities of the technique.

One of the main advantages of SANS compared to SAXS is the ability to perform contrast variation using deuterated species. This provides considerable scope to determine the structure of particular parts of a system using labelled molecules or solvent. SANS is also less destructive than SAXS can be, in some circumstances (see, e.g. the discussion of beam damage in Section 2.12), due to the lower flux and higher transmission through the sample. This chapter includes a selection of examples of the use of contrast variation in colloid and polymer sciences, as well as in the study of protein complexes in solution. SANS is also used to examine order in magnetic materials, exploiting the spin properties of neutrons (again, not possible with x-rays). This is also detailed in this chapter.

This chapter is organized as follows. Section 5.2 describes the production of neutrons, while the differential scattering cross-section and the scattering

*Small-Angle Scattering: Theory, Instrumentation, Data and Applications,*
First Edition. Ian W. Hamley.
© 2021 John Wiley & Sons Ltd. Published 2021 by John Wiley & Sons Ltd.

length of neutrons are analysed in Sections 5.3 and 5.4, respectively. Section 5.5 discusses SANS-specific data reduction considerations. Section 5.6 covers the important topic of contrast variation and gives selected examples of its application to the study of polymers including copolymer micelles, microemulsions, lipid mixtures, and protein complexes in solution. Section 5.7 outlines the theory of 'single molecule' SANS patterns that can be isolated using mixtures of deuterated and protonated molecules. Examples of SANS studies from labelled polymers including measurement of single chain properties and those of blends and block copolymers are presented in Section 5.8. Section 5.9 describes how contrast variation can be used in time-resolved studies of self-assembly of amphiphilic systems, illustrated using selected examples from the literature. Ultra-small-angle neutron scattering (USANS) is briefly outlined with selected examples in Section 5.10. This leads into a discussion of SANS and USANS (as well as SAXS and USAXS) studies on porous materials in Section 5.11. SANS studies on magnetic materials including those that exploit the neutron spin properties are discussed in Section 5.12. The relatively new and as yet undeveloped technique of Spin-echo small-angle neutron scattering (SESANS) is introduced in Section 5.13. The chapter concludes with a few lines in Section 5.14 on the use of complementary SAXS and SANS.

## 5.2   PRODUCTION OF NEUTRONS

Neutrons at reactor sources are produced by nuclear fission reactions of uranium ($^{235}$U) in a chain reaction, which is moderated using water or heavy water [1, 2].

In contrast, spallation sources generate neutrons when a proton beam (generated by a synchrotron) bombards a target that comprises a vessel of a heavy metal such as tungsten or steel filled with liquid mercury. This is an efficient process, around 30 neutrons being produced per incident proton [3]. The proton energy at currently operating spallation sources is 0.8–3 GeV with a repetition rate 20–60 Hz with 100 ns neutron bunches in the pulses [1, 2].

The different processes involved with fission production of neutrons at reactors as opposed to neutrons produced by spallation are shown in Figure 5.1.

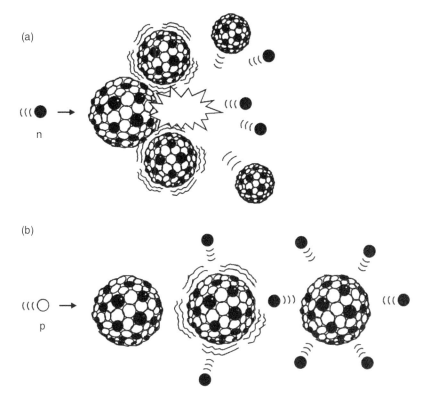

**Figure 5.1** (a) Nuclear fission reaction of $^{235}$U nuclei (shown as balls of protons and neutrons) involving neutrons, (b) spallation is a process in which a beam of protons bombards a heavy metal target producing neutrons.

## 5.3 DIFFERENTIAL SCATTERING CROSS-SECTION

The measured intensity of neutrons (in cm$^{-1}$) scattered per solid-angle element $\Delta\Omega$ of a detector during a time $\Delta t$ can be written as [4–6]

$$I(q) = \frac{d\sigma}{d\Omega}(q)\Phi_0\Delta\Omega\Delta t \frac{I_{0,s}}{I_{0,i}} at\varepsilon(\lambda) \tag{5.1}$$

Here, $\frac{d\sigma}{d\Omega}(q)$ is the differential scattering cross-section, $\Phi_0$ is the incident flux (units: cm$^{-2}$ s$^{-1}$), $I_{0,i}/I_{0,s}$ is the transmission (ratio of the intensity before and after the sample, Section 2.4), $a$ is the beam size (illuminated area on the sample in cm), and $t$ its thickness (cm) and $\varepsilon(\lambda)$ is the detector efficiency at a given wavelength.

The incident flux of neutrons is much lower than that of x-rays, at a high-power reactor or spallation SANS beamline, a flux at the sample would typically be $\sim 10^6-10^7$ neutrons cm$^2$ s$^{-1}$ at the peak wavelength (based on quoted values for beamline CG-2 at ORNL [7], NG-7 at NIST [8], D22/D33 at the ILL [9], and beamline SANS2D at ISIS, see also Ref. [2]). The flux from the source is higher, being typically $\sim 10^{15}$ neutrons cm$^2$ s$^{-1}$ for a high-power reactor or spallation source [10]. The flux at the sample compares to a value $\sim 10^{13}$ photons s$^{-1}$ typical for third or fourth synchrotron SAXS beamlines [11–13]. Note that flux for SAXS beamlines is often quoted in photons s$^{-1}$, for a given slit collimated beam size. However, brilliance is a better measure of brightness for SAXS beamlines, as discussed in Section 4.2.

## 5.4   SCATTERING LENGTHS

In contrast to the atomic scattering factors for x-rays, for neutrons the scattering lengths are independent of $q$ and do not show any trend as a function of atomic number, since they are the result of complex nuclear processes. Table 5.1 gives values for some elements and isotopes.

For neutron scattering, especially for samples containing hydrogen, incoherent scattering is significant. This scattering arises (for a single isotope) from fluctuations in the scattering from different neutron spin states. For an element, it is also necessary to average the contributions to the scattering from the different isotopes. Considering a sample containing a single isotope, the differential scattering cross-section can be written

$$\frac{d\sigma}{d\Omega} = \sum_j \sum_k \langle b_j b_k \rangle \exp[-i\mathbf{q}.(\mathbf{r}_j - \mathbf{r}_k)] \tag{5.2}$$

where $\langle b_j b_k \rangle$ denotes a mean value of the product of scattering lengths. Separating out terms for which $j = k$ and $j \neq k$ gives, respectively:

$$\langle b_j b_j \rangle = \langle b_j^2 \rangle = \langle b^2 \rangle \tag{5.3}$$

and

$$\langle b_j b_k \rangle = \langle b_j \rangle \langle b_k \rangle = \langle b \rangle^2 \tag{5.4}$$

These equations can be combined to give

$$\langle b_j b_k \rangle = \langle b \rangle^2 + \delta_{j,k}(\langle b^2 \rangle - \langle b \rangle^2) \tag{5.5}$$

**Table 5.1**  Neutron scattering lengths and cross sections for elements/main isotopes in soft materials [14].

| Element/Isotope (abundance) | $b_{coh}$ (fm) | $b_{incoh}$ (fm) | $\sigma_{coh}$ (barn) | $\sigma_{incoh}$ (barn) |
|---|---|---|---|---|
| H | −3.7390 | — | 1.7568 | 80.26 |
| $^1$H (99.985%) | −3.7406 | 25.274 | 1.7583 | 80.27 |
| $^2$H (0.015%) | 6.671 | 4.04 | 5.592 | 2.05 |
| C | 6.6460 | — | 5.551 | 0.001 |
| $^{12}$C (98.9%) | 6.6511 | 0 | 5.559 | 0 |
| $^{13}$C (1.1%) | 6.19 | −0.52 | 4.81 | 0.034 |
| N | 9.36 | — | 11.01 | 0.5 |
| $^{14}$N (99.63%) | 9.37 | 2.0 | 11.03 | 0.5 |
| $^{15}$N (0.37%) | 6.44 | −0.22 | 5.21 | 0.00005 |
| O | 5.803 | — | 4.232 | 0.0008 |
| $^{16}$O (99.762%) | 5.803 | 0 | 4.232 | 0 |
| $^{18}$O (0.2%) | 5.84 | 0 | 4.29 | 0 |
| Al | 3.449 | 0.256 | 1.503 | 0.0082 |
| P | 5.13 | 0.12 | 3.307 | 0.005 |
| S | 2.847 | — | 1.0186 | 0 |
| $^{32}$S (95.02%) | 2.804 | 0 | 0.988 | 0 |
| $^{34}$S (4.21%) | 3.48 | 0 | 1.52 | 0 |

*Source:* Data from Ref. [14].

where $\delta_{j,k}$ is the Kronecker delta, which takes the value $\delta_{j,k} = 1$ if $j = k$; otherwise $\delta_{j,k} = 0$. Substitution of Eq. (5.5) into Eq. (5.2) gives

$$\frac{d\sigma}{d\Omega} = \langle b \rangle^2 \sum_j \sum_k \exp[-i q.(\mathbf{r}_j - \mathbf{r}_k)] + N(\langle b^2 \rangle - \langle b \rangle^2) \quad (5.6)$$

The first term on the righthand side of this equation is due to the coherent scattering. The second represents incoherent scattering due to fluctuations in the scattering length. The coherent and incoherent scattering lengths are then written as

$$b_{coh} = \langle b \rangle \quad (5.7)$$

and

$$b_{incoh} = (\langle b^2 \rangle - \langle b \rangle^2)^{1/2} \quad (5.8)$$

The scattering cross sections (total scattered in all directions) are then given by

$$\sigma_{coh} = 4\pi \langle b \rangle^2 \quad (5.9)$$

$$\sigma_{incoh} = 4\pi(\langle b^2 \rangle - \langle b \rangle^2) \quad (5.10)$$

Inspection of Table 5.1 shows that the incoherent scattering is particularly large for hydrogen, which leads to a large flat 'background' scattering from all hydrogenous materials. This can be avoided where possible by deuteration of the sample or use of deuterated solvent for studies on solutions.

Neutron scattering lengths in SI units are given in units of $\times 10^{-15}$ m (fm), although cm or Å are widely used in practice. Scattering cross sections are conventionally given in non-SI units of $\times 10^{-24}$ cm$^2$ (barn). The values in Table 5.1 are taken from Ref. [14], an alternative source (with slightly different values) is the Neutron Data Booklet [5].

The scattering contrast in a two-component system is the difference in their scattering length densities. The scattering length density (SLD) of a molecule is defined as

$$\rho = \frac{\sum_{i=1}^{N} b_{\text{coh},i}}{V_m} \tag{5.11}$$

Here the sum is over the $N$ atoms in a molecule and $V_m$ is its molar volume. The values of SLD are usually given in Å$^{-2}$ or cm$^{-2}$.

## 5.5   SANS DATA REDUCTION CONSIDERATIONS

General reduction methods for SAS data are the subject of Chapter 2, but further details are provided here on resolution function analysis, which is particularly important for SANS data modelling, along with a brief discussion of multiple scattering correction, which can be more significant for SANS data than for SAXS since the sample is often thicker. Inelastic scattering effects are also briefly mentioned.

### 5.5.1   Resolution Function

The resolution function describes the finite width of scattering peaks that results from the spread in wavelength, collimation effects (beam divergence and width), sample thickness, and the detector resolution. The resolution $R(q,<q>)$ function leads to a smearing of the scattering cross section represented by the equation [15, 16]:

$$I(\langle q \rangle) = \int R(q, \langle q \rangle) \frac{d\sigma(q)}{d\Omega} dq \tag{5.12}$$

Due to the large wavelength spread $\Delta\lambda/\lambda \sim 10\%$ or more at SANS beamlines, consideration of this effect is important when analysing SANS data. For SAXS experiments, especially on synchrotron beamlines, the resolution function is much narrower and in most cases can be neglected. It is only important in special cases such as when analysing fine details of Bragg peak shape in high-resolution SAXS experiments. The SANS resolution function can be described as a Gaussian function containing terms that represent the different contributions [15], although the wavelength spread is typically dominant. The smearing effect due to sample thickness is usually small.

### 5.5.2   Multiple Scattering

This artefact, arising from the scattering of neutrons by more than one nucleus can be minimized by reducing the sample thickness. Multiple scattering may be problematic in a sample with low transmission, so this should be carefully monitored. In principle, the effects of multiple scattering can be modelled [4, 17, 18], although this is not straightforward and is not routinely done.

### 5.5.3   Inelastic Scattering

Inelastic scattering (where there is energy transfer of the neutrons, i.e. a wavelength shift) can be observed in time-of-flight SANS measurements (Section 3.6) for hydrogenous samples studied by SANS with cold neutrons with a wavelength in the range $4.5\,\text{Å} < \lambda < 20\,\text{Å}$ [4]. For example, the thermal energy of a water molecule at room temperature is much higher than the kinetic energy of a neutron and so there is inelastic energy transfer during the scattering process. This can cause, for example, the development of inelastic scattering peaks, at low scattered wavelengths (1–2 Å). This must be considered when treating data using water as a calibration standard [4]. However, in general, inelastic scattering is not problematic for SANS, since the elastically scattered neutron component can be isolated in the time-of-flight analysis.

## 5.6   CONTRAST VARIATION

### 5.6.1   Introduction

Contrast variation is a powerful method to determine features of the structure of soft materials and biomolecules by use of deuteration of the molecule

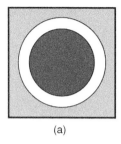 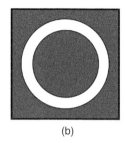 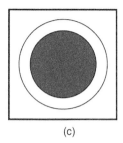

(a)                                    (b)                                    (c)

**Figure 5.2**   Schematic illustrating the contrast matching technique in small-angle neutron scattering experiments. (a) No contrast matching, (b) core contrast matching, (c) shell contrast matching.

and/or of the solvent. Labelling by selective deuteration labelling can be synthetically convenient, depending on the molecule and the sites and extent of deuteration. A variety of deuterated solvents, including $D_2O$ and deuterated organic solvents, are commercially available so contrast variation by using such solvents or mixtures with protonated solvents may readily be employed in many cases. Figure 5.2 illustrates the concepts of solvent contrast variation via changes in solvent SLD and contrast matching for the illustrative case of a core–shell micelle or colloid particle or protein. The idea applies to other particles or subcomponents of a material such as a polymer melt, and it should be borne in mind that the material (particle, polymer, etc. or regions thereof) can be deuterium labelled (as well as the solvent) to provide contrast variation. Contrast match conditions are achieved when the SLDs of two components of the system are the same. Figure 5.2a shows the case of a noncontrast matched system where the SLD of the solvent (light grey) differs from that of both core (dark grey) and shell (white). In Figure 5.2b the contrast of the solvent is matched to the core of the particle, which by Babinet's principle leads to scattering only from the shell. In Figure 5.2c, the contrast of the solvent is varied to match the shell of the particle, resulting in scattering only from the particle core.

Contrast matching for aqueous systems is achieved using $H_2O/D_2O$ mixtures. Figure 5.3 can be used to determine typical contrast matching points for different biomolecules (using typical SLD values) as a function of $D_2O$ content in the mixed solvent.

The process of H/D isotope exchange is slow; therefore, $H_2O/D_2O$ mixtures (with or without sample) should be well equilibrated before measurement. A period of at least 24 hours is recommended since most water protons and solvent-exposed protons on proteins, for example, have exchanged during this time frame [12].

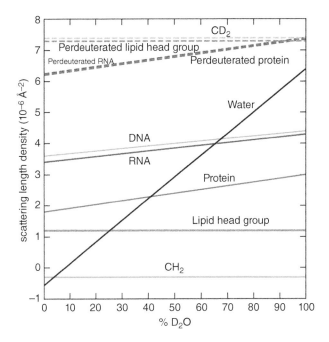

**Figure 5.3** Variation in scattering length density with $D_2O$ content for different materials [19]. The SLD of biomolecules changes with $D_2O$ content (positive slope of lines for proteins and nucleotides) due to H/D exchange. Match points correspond to the points where the lines for biomolecules cross the black line for $D_2O/H_2O$ mixture contrast. *Source:* From Zaccai et al. [19]. © 2016, Elsevier.

Since the scattered intensity is proportional to $(\Delta\rho)^2$ (as for example in Eqs. (1.6) and (1.10), see also Eq. (5.17) below), a plot of the square root of the intensity (at $q = 0$) versus the scattering density $\rho_s$ of the solvent for a mixed hydrogenated/deuterated system for example will enable determination of the contrast match point, as shown in Figure 5.4 [3, 21].

It should be noted that deuterium isotope effects can be significant, and substantial differences in melting temperatures, nonideal phase behaviour, and shifts in phase separation boundaries etc. have been observed when comparing deuterated and protonated systems, in particular in SANS studies on homopolymers [22–26] and copolymers [27]. Phase boundary shifts have also been detected via light scattering and melting/boiling point measurements, etc. Therefore, it cannot automatically be assumed that deuteration does not change the physical properties of the system.

Although it is impossible to review all applications of contrast variation SANS in the literature within the space available, selected examples are

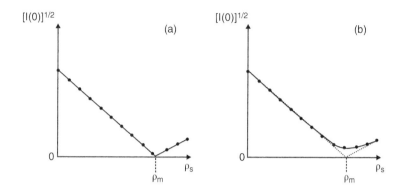

**Figure 5.4**  A plot of the square root of the intensity versus the solvent scattering length density for a contrast variation series (e.g. series of hydrogenous and deuterated solvent mixtures) can be used to define the contrast matching point (solvent scattering length density $\rho_m$ at this point). (a) Ideal case of perfectly homogeneous system, (b) system with inhomogeneities in particle scattering. *Source:* Redrawn from Williams [20].

given in the following from several different classes of material, including polymers, microemulsions, lipid mixtures, and proteins and protein complexes.

## 5.6.2   Contrast Variation Studies on Polymers

In an early example of a study using selective deuteration of selected parts of polymer chains to provide conformational data on the labelled region, Matsushita et al. investigated the conformation of PS-*b*-P2VP [polystyrene-*b*-poly(2-vinyl pyridine)] with deuterated PS sequences at the end of the PS block or adjacent to the junction with the P2VP block (the other part of the PS chain being undeuterated) [28]. An aligned sample of the block copolymer (in the glassy state), which formed a lamellar phase, was studied in order to examine the conformation perpendicular to the lamellar interface. The differential contrast provided in SANS, along with combined SAXS data, enabled detailed information on the scattering density profiles normal to the lamellar interface to be obtained. This provided the conformations shown schematically in Figure 5.5. The junction labelled block was found to be adjacent to the interface, with a conformation compressed perpendicular to the lamellar normal. In contrast, the end deuterated PS block was located in the centre of the PS domain, with an unperturbed conformation [28].

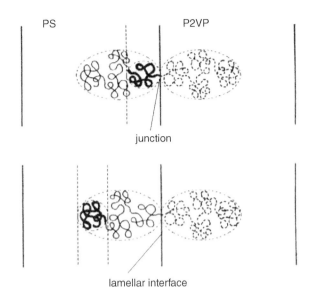

**Figure 5.5** Schematic showing conformation of different regions of the PS chain (deuterated regions shown darkened) with respect to the lamellar interface in a PS-*b*-P2VP diblock copolymer, deduced from combined SANS and SAXS on an aligned specimen. *Source:* Adapted from Matsushita et al. [28].

In another study, a contrast variation series proved vital in constraining the fitting of form factors of block copolymer micelles, which provided unique information on the micelle structure and chain packing [29]. The system was a PS-*b*-PI [PI: polyisoprene] diblock copolymer in di-*n*-butyl phthalate (DBP), which is a selective solvent for the PS block, leading to micelles with a PI core. Figure 5.6 shows a contrast variation series of SANS measurements using mixtures of deuterated and normal DBP, combined with an additional contrast provided by SAXS data. This set of data was fitted simultaneously to different models of the micelle shape, allowing for the core and attached chains, including the sphere with attached Gaussian chains model in Table 1.2 [30, 31]. Analogous expressions for ellipsoids and cylindrical micelles applied to model the form factors of block copolymer micelles were also used. The best overall fit was obtained for a model of polydisperse cylindrical micelles with attached chains [29].

An example of the use of SANS to investigate core–shell particles (cf. Figure 5.2) is a study on triblock copolymer micelles in $H_2O/D_2O$ solutions [32]. Contrast matching was used for the outer water-soluble poly(acrylic acid) [PAA] block and a different $H_2O/D_2O$ composition led to matching of the PMMA [poly(methyl methacrylate)] midblock in

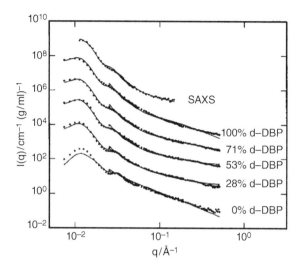

**Figure 5.6** Contrast variation series of SANS measurements (combined with an additional SAXS contrast) for a PS-*b*-PI diblock in DBP solvent mixtures (deuterated DBP contents indicated) along with simultaneously fitted form factor data, using a form factor of a cylinder with attached chains. *Source:* From Pedersen et al. [29]. © 2000, American Chemical Society.

a PEHA-*b*-PMMA-*b*-PAA [PEHA: poly(2-ethylhexyl acrylate)] triblock, forming micelles with a hydrophobic core and inner shell and a hydrophilic outer shell of PAA (Figure 5.7). These measurements permitted the determination of the radius of the core and first shell (in the case of outer shell contrast matching) and of the core and outer shell (in the case of inner shell contrast matching), as shown in Figure 5.7b.

SANS from block copolymer melts is discussed in brief in Section 5.8, and further information on the quite extensive SANS and SAXS studies of block copolymers in the melt and solutions states is available elsewhere [31, 33–36].

## 5.6.3   Contrast Variation Studies on Microemulsions

Contrast variation SANS has been used to elucidate the structure of microemulsions. For example, a study of bicontinuous microemulsions formed by the surfactant sodium bis(ethylhexyl) sulfosuccinate (AOT) in oil(decane)/water examined the transition from a water-in-oil microemulsion to an oil-in-water microemulsion on heating, via a bicontinuous

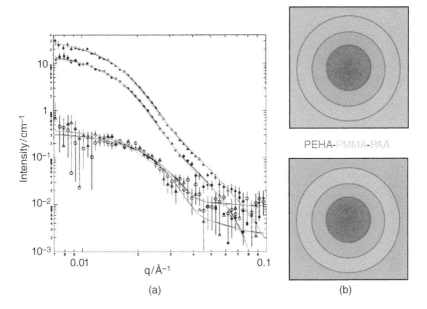

**Figure 5.7** (a) Measured SANS (symbols) data with fits to sphere models (lines). Upper two curves: contrast matched PAA corona in acid form, 49% $D_2O$ (●) and neutralized form, 63% $D_2O$ (▲). Bottom curves: zero contrast for PMMA (24% $D_2O$, acid form (o), neutralized form (Δ)). *Source*: Adapted from Ref. [32]. (b) Colour-coded contrast scheme for the two data sets with the solvent contrast varied to match PAA (top: green) or PMMA (bottom: blue).

microemulsion, a so-called structural inversion [37]. The SANS measurements were performed with nondeuterated decane, providing 'bulk contrast' (between oil and water) and with deuterated decane, producing 'interface contrast' (between the surfactant and the oil and water). Figure 5.8 shows measured SANS data at several temperatures, along with fits to the Teubner-Strey structure factor (Section 1.6.6). The peak position and Teubner-Strey parameter $1/k\xi$ show different behaviour above and below 40 °C, which was interpreted as the inversion temperature. The data were also analysed using a modified Berk model (Section 1.6.6), as well as the Teubner-Strey equation.

## 5.6.4   Contrast Variation Studies on Lipid Mixtures

One of the first studies to examine phase separation in lipid mixtures by SANS used contrast variation with mixtures of deuterated and non-deuterated lipids at fixed $H_2O/D_2O$ content (this was termed

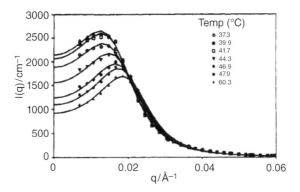

**Figure 5.8** SANS data at the temperatures indicated (symbols) along with Teubner-Strey structure factor fits (lines) for microemulsions (under bulk contrast) formed by surfactant AOT-water decane as described in the text. *Source:* From Chen et al. [37]. © 1991, American Chemical Society.

'inverse contrast variation') [38, 39]. The lipid mixture studied was 1,2-dimyristoyl-*sn*-glycero-3-phosphocholine (DMPC) with cholesterol. A mixture of perdeuterated and protonated DMPC was used. Figure 5.9 shows an example of a plot of $\sqrt{I(0)}$ versus $x$, the mole fraction of deuterated DMPC (this plot is similar to that shown in Figure 5.4) [38]. The intensity at $q = 0$, $I(0)$ was obtained from a Kratky-Porod plot using Eq. (2.23). The data shown in Figure 5.9 could be fitted to obtain the mole fractions of cholesterol in the two phases in the phase separated system (in this particular case $x_c = 0.08$ and $x_c = 0.24$) [38]. This group also report standard $H_2O/D_2O$ contrast variation studies for the same lipid and various mixtures [38–40].

## 5.6.5   Contrast Variation Studies of Proteins and Protein Complexes

Contrast variation SANS has been used to investigate the structure of proteins and of protein complexes. A good example used selective deuteration of a protein that forms a complex with a (nondeuterated) inhibitor [41]. The study involved a histidine kinase (KinA) and a DNA damage checkpoint inhibitor (Sda). Solution SAXS data was measured along with a SANS contrast variation series using deuterated Sda and nondeuterated KinA in a series of $H_2O/D_2O$ mixtures. SAXS showed that KinA forms dimers in solution, as does Sda, therefore, the complex is represented as $KinA_2\text{-}2^D Sda$. The combined data for the complex was first analysed to provide $R_g$ and $p(r)$.

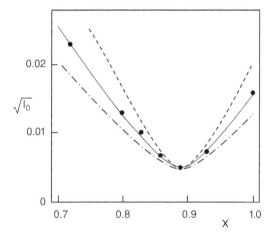

**Figure 5.9** Variation of $\sqrt{I_0}$ with $x$, the mole fraction of deuterated DMPC in the mixture DMPC(1-$x$): DMPC-$d54$($x$) with cholesterol (10%), at 7 °C. The lines were calculated according to a simple model for a phase separated mixture. The solid line is the best fit to the data (solid points), the other two lines were used to estimate uncertainties on the determined compositions of the two phases. *Source:* Redrawn from Knoll et al. [38].

The dataset was also fitted using the rigid body simulated annealing ATSAS routine SASREF (Section 4.7). It was found that although Sda molecules in solution exist as dimers, they bind to KinA$_2$ as monomers, binding to the base of the KinA dimer and suggesting a mechanism for the inhibition of KinA auto-kinase activity. Part of the analysis involved determination of the radius of gyration of the complex using a Sturhmann plot (Figure 5.10), according to the quadratic equation [42]:

$$R_g^2 = R_m^2 + \frac{\alpha}{\Delta\rho} + \frac{\beta}{(\Delta\rho)^2} \tag{5.13}$$

Here, $R_m$ is the radiation of gyration at infinite contrast, $\overline{\Delta\rho}$ is the mean contrast for the complex (the volume fraction weighted sum of the contrasts for the two components) and $\alpha$ and $\beta$ are scattering density related coefficients obtained from the fit (Figure 5.10). The sign of $\alpha$ defines whether the centre of mass of the higher- or lower-scattering density component is closer to the centre of mass of the complex, and the magnitude of $\beta$ is related to the separation of the centres of mass of the two components [19, 43, 44]. Contrast variation SANS studies of protein complexes have been briefly reviewed [45]. A linear plot is obtained in the case that $\beta = 0$.

The parallel axis theorem can be used to calculate the radius of gyration for a two-component system with a large difference in SLDs for the two

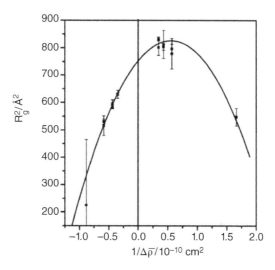

**Figure 5.10** Sturhmann plot for radius of gyration (squared) of a $KinA_2$-$2^D$Sda complex versus the mean contrast for the complex [41]. *Source:* From Whitten et al. [41]. © 2007, Elsevier.

components [19, 44, 46]:

$$R_g^2 = F_1 R_1^2 + F_2 R_2^2 + F_1 F_2 D_{cm}^2 \qquad (5.14)$$

Here, $R_1$ and $R_2$ are the radii of gyration of the components, $D_{cm}$ is the centre-of-mass separation and the scattering coefficients are

$$F_i = \frac{\Delta \rho_i V_i}{\Delta \rho V} \qquad (5.15)$$

Here $\Delta \rho_i$ and $V_i$ refer, respectively, to the scattering density contrast and molar volume of component $i$, and $\Delta \rho$ and $V$ are the corresponding quantities for the complex.

Contrast variation SANS was used to determine the structure of a complex of a chaperone protein Skp with two unfolded outer membrane (uOMP) proteins, OmpA and OmpW [19]. To create contrast, the uOMPs were expressed by cells grown in partially deuterated media, leading to replacement of nonexchangeable hydrogen atoms, i.e. those bound to carbon, with deuterium. This was not done for Skp, which was expressed by cells in undeuterated media. After expression, the Skp was prepared in $H_2O$- or $D_2O$-based buffers. The contrast matching point for the proteins in the complex was determined by plotting measured $\sqrt{I(0)/c}$ vs. $f_{D_2O}$, the mass fraction of $D_2O$. This plot follows from the expression (from

Eq. (2.10)) for the forward scattering for a two-component system with two different contrasts [19]:

$$I(0) = n_p(\Delta\rho)^2 V^2 = n_p[f_1\Delta\rho_1 V_1 + f_2\Delta\rho_2 V_2]^2 \quad (5.16)$$

Here, $n_p$ is the number density of protein complexes, $f_1$ and $f_2$ denote the mass fractions of the two components, $\Delta\rho_1$ and $\Delta\rho_2$ are their SLD contrasts (with respect to solvent), and $V_1$ and $V_2$ are their volumes. Since $n_p = cN_A/M$, where $c$ is the concentration of protein complex (in g cm$^{-3}$), $N_A$ is Avogadro's number and $M$ is the molar mass in g mol$^{-1}$, Eq. (5.16) gives [21]

$$\sqrt{I(0)/c} = \sqrt{N_A/M}[f_1\Delta\rho_1 V_1 + f_2\Delta\rho_2 V_2] \quad (5.17)$$

Plots of $\sqrt{I(0)/c}$ vs. $f_{D_2O}$ are linear [12, 21], and the intercepts $\sqrt{I(0)/c} = 0$ give the contrast match points (a deviation from linearity is observed if the system is heterogeneous [12]). Note that, distinct from Figure 5.4, these authors have taken the negative root in Eq. (5.17) above the match point in order to have a straight line over the whole range of $f_{D_2O}$ (Figure 5.11) The

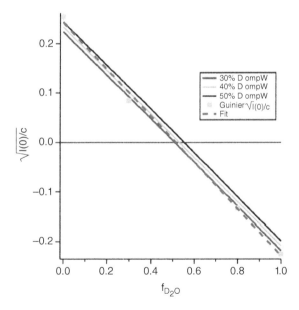

**Figure 5.11** Determination of contrast match point from SANS experiments on a two-component protein complex [19]. Data are presented with $I(0)$ from Guinier plots of measured data at several solvent contrasts (with linear fit to the data points) along with calculated curves, assuming different levels of deuteration of the ompW (grown in partly deuterated media). The match point is at $(51 \pm 2)\%$ D$_2$O. *Source:* From Zaccai et al. [19]. © 2016, Elsevier.

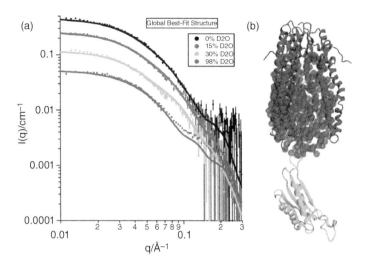

**Figure 5.12** (a) SANS data at several contrasts along with fits based on a best model, shown in (b). The blue and green helices show Skp, and the red ellipsoids represent OompA [19]. The blue helices represent the Skp transmembrane domain and the green helices are the Skp periplasmic domain, which folds independently. *Source:* From Zaccai et al. [19]. © 2016, Elsevier.

plot shows data with $I(0)$ values obtained from Guinier plots (Eq. (1.24)), along with lines calculated allowing for different levels of deuteration of the uOMP components (the line calculated for 30% deuteration matches the measured data points most closely).

The theoretical contrasts used in the plots in Figure 5.11 were calculated using the SLD contrast calculator available in SASSIE (Table 2.2) [47]. From a simultaneous fit of the SANS data set measured for several $H_2O/D_2O$ contrasts, models for the Skp-uOMP complexes were obtained [19]. Figure 5.12 shows the fitted SANS data and best fit model structure for Skp-OmpA. The models were obtained via MD and Monte Carlo simulations in SASSIE (Table 2.2) with constraints from NMR and crystal structures (using an available crystal structure for Skp). The model structure shows the 'tentacles' of the Skp structure wrapping the OmpA, with the tips of the tentacles around the transmembrane domain.

Another example of the use of $H_2O/D_2O$ solvent contrast variation applied to protein-based assemblies is a study of the structure of lipoprotein particles of two types, low- and high-density lipoproteins (LDLs and HDLs) [48]. These particles are present in blood plasma, and their levels are associated with cardiovascular conditions. Three contrast conditions, 5%, 42%, and 100% $D_2O$ were selected. In the first case, the lipid is contrast

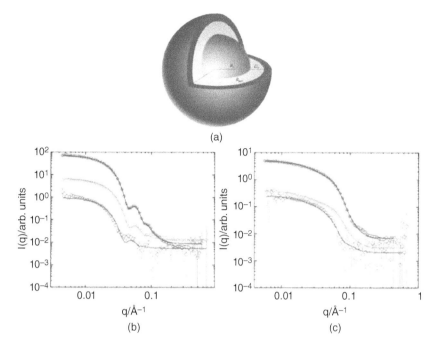

**Figure 5.13** SANS data for contrast matched lipoprotein particles. (a) Schematic of particle, comprising a core of fats (triglycerides, cholesterol, and cholesteryl ester), (b) SANS data (open circles) and form factor fits (solid lines) for LDL particles, (c) SANS data (open circles) and form factor fits (solid lines) for HDL particles. Blue: 100% $D_2O$, yellow: 42% $D_2O$, orange: 5% $D_2O$. *Source:* From Maric et al. [48]. Licensed under CC BY 4.0.

matched to the solvent, in the second condition the protein is contrast matched and in 100% $D_2O$ scattering from both lipid and protein is observed. Figure 5.13 shows the corresponding SANS data measured, along with fits to model form factors of core-shell spheres [48]. This enabled the determination of the total and core thickness of the particles, as well as the thickness of the inner and outer shells, corresponding respectively to the phospholipid acyl chains and phospholipid head groups and protein (yellow and blue shells in Figure 5.13).

## 5.7 SINGLE MOLECULE SCATTERING FROM MIXTURES OF PROTONATED AND DEUTERATED MOLECULES

By using mixtures of protonated and deuterated molecules, it is possible to obtain solely intramolecular scattering, even from polymer melts or thermotropic liquid crystals where structure factors would normally obscure the intramolecular scattering. This can be achieved using a random mixture of labelled and unlabelled molecules, it does not require use of a dilute mixture of deuterated molecules. The following gives a derivation of the equations that shows how the single molecule scattering from a difference molecule (with a SLD difference between protonated and deuterated molecules) is obtained. Variations of the following derivation can be found in the literature [49–55]. We start from a general expression for the intensity written (cf. Eq. (1.4)) as

$$I(\mathbf{q}) = \int A(\mathbf{q})A^*(\mathbf{q}) \exp[-i\mathbf{q}.\mathbf{r}]d\mathbf{r} \qquad (5.18)$$

Here $A(\mathbf{q})$ is the amplitude form factor (Eq. (1.2)) for a given molecule/particle. To separate intra- and intermolecular terms, we introduce a conditional pair distribution function [54]:

$$P(\mathbf{r}) = \delta(\mathbf{r}) + P_D(\mathbf{r}) \qquad (5.19)$$

where $\delta(\mathbf{r})$ is the Kronecker delta function (which takes the value $\delta(\mathbf{r}) = 1$ for the reference molecule at $\mathbf{r} = 0$ or $\delta(\mathbf{r}) = 0$ otherwise) and $P_D(\mathbf{r})$ is the conditional probability of finding a different molecule at $\mathbf{r} \neq 0$. This leads to [54]

$$I(\mathbf{q}) = \int (\delta(\mathbf{r}) + P_D(\mathbf{r}))A(\mathbf{q})A^*(\mathbf{q}) \exp[-i\mathbf{q}.\mathbf{r}]d\mathbf{r} \qquad (5.20)$$

Now, we consider a mixture of two molecules labelled 1 and 2 for generality, although to be concrete and for most applications, these can be considered to be H and D labelled molecules. The corresponding volume fractions are denoted $\phi_1$ and $\phi_2$. Considering that there are four possible combinations of molecules at the origin and at $\mathbf{r}$, and assuming random mixing, the intensity can be written as [54]

$$I(\mathbf{q}) = \int \{\phi_1(\delta(\mathbf{r}) + \phi_1 P_D(\mathbf{r}))A_1(\mathbf{q})A_1^*(\mathbf{q}) + \phi_1\phi_2 P_D(\mathbf{r})A_1(\mathbf{q})A_2^*(\mathbf{q})$$

$$+ \phi_1\phi_2 P_D(\mathbf{r})A_2(\mathbf{q})A_1^*(\mathbf{q}) + \phi_2(\delta(\mathbf{r}) + \phi_2 P_D(\mathbf{r}))A_2(\mathbf{q})A_2^*(\mathbf{q})\}$$

$$\exp[-i\mathbf{q}.\mathbf{r}]d\mathbf{r} \qquad (5.21)$$

The terms in the integral can be rearranged to collect 'pure' and 'mixture' components, i.e.

$$I(\mathbf{q}) = \int \{\delta(\mathbf{r})\phi_1\phi_2(A_1(\mathbf{q}) - A_2(\mathbf{q}))(A_1^*(\mathbf{q}) - A_2^*(\mathbf{q})) + (\delta(\mathbf{r}) + P_D(\mathbf{r}))(\phi_1 A_1(\mathbf{q})$$

$$+ \phi_2 A_2(\mathbf{q}))(\phi_1 A_1^*(\mathbf{q}) + \phi_2 A_2^*(\mathbf{q}))\} \exp[-i\mathbf{q}.\mathbf{r}]d\mathbf{r} \qquad (5.22)$$

This gives rise to the 'pure' and 'mixture' terms in $I(q) = I_{pure}(q) + I_{mix}(q)$:

$$I_{pure}(\mathbf{q}) = \int \{(\phi_1 A_1(\mathbf{q}) + \phi_2 A_2(\mathbf{q}))(\phi_1 A_1^*(\mathbf{q}) + \phi_2 A_2^*(\mathbf{q}))\} P(\mathbf{r}) \exp[-i\mathbf{q}.\mathbf{r}] d\mathbf{r} \tag{5.23}$$

and

$$I_{mix}(\mathbf{q}) = \phi_1 \phi_2 \int \{(A_1(\mathbf{q}) - A_2(\mathbf{q}))(A_1^*(\mathbf{q}) - A_2^*(\mathbf{q}))\} \exp[-i\mathbf{q}.\mathbf{r}] d\mathbf{r} \tag{5.24}$$

Since the scattering from a pure liquid is very small at low $q$ because it is related to the structure factor, which is small at low $q$, the total (coherent) intensity is dominated by the $I_{mix}$ term, which corresponds to the scattering from the 'difference' molecule, with the scattering factors in the equations for the amplitude factors (Eq. (1.2)) replaced by those for the difference (in general $b_H - b_D$ for each H/D isotopic substitution in the molecule).

For mixtures of protonated (volume fraction $\phi_1 = \phi$) and deuterated (volume fraction $\phi_2 = 1 - \phi$) polymers, the isotropic version of Eq. (5.24) is commonly written in the following form [3, 49, 52, 53, 55]:

$$I_{mix}(q) = \phi(1 - \phi)(b_H - b_D)^2 N n_{seg}^2 P(q) \tag{5.25}$$

Here $N$ is the number of polymer molecules per unit volume, $n_{seg}$ is the number of segments per chain, and $P(q)$ is the form factor, here in dimensionless form, for example the Debye function in Eq. (1.107).

As well as investigation of the conformation of polymer chains in mixtures of deuterated and protonated macromolecules, this theory has been used to investigate the ordering of small molecules, for example the single molecule orientational ordering of liquid crystals [54, 56, 57].

## 5.8   SANS FROM LABELLED POLYMERS

SANS has been widely used to investigate the conformation of many types of polymers, using mixtures of deuterated and protonated chains. SANS measurements on mixtures of deuterated and protonated versions of the same polymer have enabled measurement of the dimensions of polymer chains in the melt and in solution that has provided unique insights into the understanding of polymer conformation. SANS has also been used on blends of different polymers, one or both of which can be labelled to improve contrast and/or reduce incoherent scattering, which is high for protonated materials (Section 5.4). Again, SANS has provided some of the key data underpinning the understanding of the thermodynamics of polymer blend miscibility or phase separation. In similar ways, SANS has been used to examine the conformation of copolymer chains in the melt and solution and inter-molecular interactions through structure factor measurements. The purpose of this section is not to review the large body of literature on this topic (which is the subject of many papers, reviews [58, 59] and dedicated books [3, 55]), but to provide a brief account, with examples, of the essential features of SANS measurements from polymer melts, blends, and copolymers.

The first examples are of the determination of polymer chain conformation in the melt and in dilute solution. As discussed in Section 1.8, polymer chains in the melt should take dimensions of Gaussian (random) coils. The form factor should be described by the Debye function Eq. (1.107) with a scaling of intensity given by Eq. (1.108). Representative data that supports this analysis is provided by SANS intensity profiles measured for mixtures of deuterated and protonated polystyrene at several compositions (Figure 5.14). Also, according to Eq. (5.25), the intensity at $q = 0$ should be proportional to $\phi(1 - \phi)$. This is confirmed in the inset of Figure 5.14.

The next example shows clear evidence for the intensity scaling expected for excluded volume interactions of polymer chains in solution (Section 1.8), $I(q) \sim q^{-1/0.588} = q^{-1.7}$. This scaling is exhibited for semiflexible polymer chains in dilute and semidilute solution in a good solvent, as illustrated by the data in Figure 5.15 for polystyrene in deuterated toluene [61] .

In the same study [61], the structure factor effects evident in the data in Figure 5.15 were analysed using Monte Carlo simulations of chain

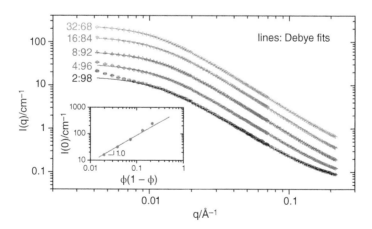

**Figure 5.14** Measured SANS intensity profiles (symbols) for mixtures of deuterated (molar mass $M_w = 253 \, \text{kg mol}^{-1}$) and protonated ($M_w = 222 \, \text{kg mol}^{-1}$) polystyrene at the indicated H:D volume fraction ratios fitted with Debye functions (lines), inset: scaling of $I(0)$ with $\phi(1 - \phi)$. *Source*: From Wang et al. [60]. © 2020, Elsevier.

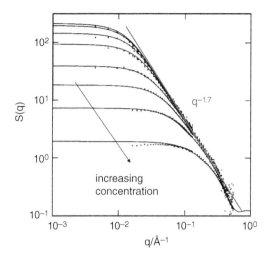

**Figure 5.15** SANS data for polystyrene with $M_w = 120\,\mathrm{kg\,mol^{-1}}$ in deuterated toluene in the concentration range $c = 0.0013$ to $0.27\,\mathrm{g\,ml^{-1}}$. The scaling of intensity with $q^{-1.7}$ at high $q$ is highlighted. The data span the dilute and semidilute regimes and the effect of increasing structure factor is evident. The points are measured data and the lines are calculated from Monte Carlo simulations of semiflexible chains. *Source:* Adapted from Pedersen and Schurtenberger [61].

conformations and using the polymer reference interaction site (PRISM) model expression for the structure factor:

$$S_{\mathrm{PRISM}}(q) = \frac{P(q)}{1 + \beta C(q)P(q)} \qquad (5.26)$$

Here $P(q)$ is the form factor (of a single chain), $C(q)$ is the normalized Fourier transform of the direct correlation function (cf. Eq. (1.75)) for the spheres on the chains and $\beta$ is an adjustable parameter related to compressibility.

For polymer blends, SANS can be used to determine the Flory-Huggins interaction parameter $\chi$ and, by extrapolation, the spinodal temperature for phase separation. The spinodal temperature can be obtained from the structure factor at $q = 0$, $S(0)$, which by Eq. (1.110) is related to the second derivative of the free energy, which defines the spinodal when $S(0) = 0$. To determine $S(0)$, the inverse of Eq. (1.117) may be used:

$$\frac{1}{S(q)} = \frac{1}{S(0)}(1 + q^2\xi^2) \qquad (5.27)$$

This shows that a plot of $S(q)^{-1}$ versus $q^2$ gives $S(0)$ from the intercept (and the correlation length $\xi$ from the gradient). Figure 5.16 gives an example of

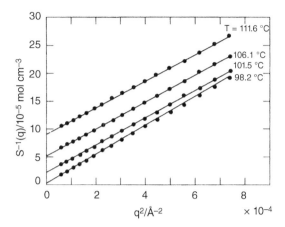

**Figure 5.16** Linear dependence of $S(q)^{-1}$ versus $q^2$ at the temperatures indicated for a blend of deuterated polybutadiene ($d$-PB) and polystyrene (PS) with degrees of polymerization $n$(d-PB) = 29 and $n$(PS) = 19 with a volume fraction of PS, $\phi = 0.53$. *Source:* From Janssen et al. [62]. © 1992, American Physical Society.

Ornstein-Zernike plots according to Eq. (5.27) for a blend of deuterated polybutadiene and polystyrene at several temperatures.

According to mean field theory (random phase approximation), from Eq. (1.115), $1/S(0)$ should be a linear function of $1/T$ since (empirically) the Flory-Huggins interaction parameter is inversely proportional to temperature: $\chi = A/T + B$. This behaviour is exemplified by the data in Figure 5.17 at high temperature where $S^{-1}(q)$ decays linearly with $T^{-1}$. The slope of this plot is negative; however, for some phase-separating blends such as the system for which data are presented in Figure 5.16, the slope can be positive (according to the temperature-dependence of $\chi$).

The data in Figure 5.17 shows curvature as temperature is lowered towards the phase-separation boundary. This is expressed by writing [62–64]

$$\frac{1}{S(0)} = \left(\frac{T_c - T}{T_c}\right)^\gamma \tag{5.28}$$

where $T_c$ is the critical temperature (at which $S^{-1}(0) = 0$) and $\gamma = 1$ in the mean field case but close to the transition there is a deviation from this behaviour and the data in Figure 5.17 are described by $\gamma = 1.26$, which is the value of the exponent for the Ising model. The extrapolated mean field critical temperature (spinodal temperature) is denoted $T_C^{MF}$, while the actual measured value from the SANS data is $T_c$.

In a similar fashion, the apparent spinodal temperature of a block copolymer melt may be obtained from a plot of $S(q)^{-1}$ versus $1/T$ (*Note:* For block

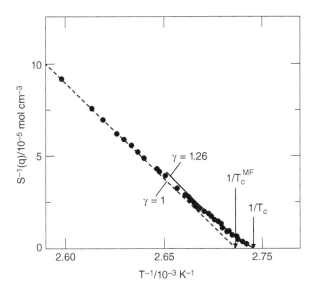

**Figure 5.17** Plot of the inverse structure factor obtained from SANS measurements versus inverse temperature for a blend of deuterated polybutadiene (d-PB) and polystyrene (PS) with degrees of polymerization $n(d\text{-PB}) = 29$ and $n(\text{PS}) = 19$ with a volume fraction of PS, $\phi = 0.53$. The dashed straight line is consistent with mean field theory (Eq. (5.28) with $\gamma = 1$). The data show deviations from mean field behaviour close to the critical temperature $T_c$ with curvature described by Eq. (5.28) with $\gamma = 1.26$, the exponent for the Ising model. These data show the mean field-to-Ising crossover. *Source:* From Janssen et al. [62]. © 1992, American Physical Society.

copolymers, the analysis is performed at $q = q^*$, the position of the observed peak in the SANS intensity profile due to concentration fluctuations). This is consistent with the expression 1.120. However, for a block copolymer, the order–disorder or (in older nomenclature) microphase separation transition occurs at a lower temperature, and this leads to nonlinear behaviour before the spinodal temperature can be reached, again due to fluctuation effects, discussed in detail elsewhere [33, 65–67]. This is illustrated in Figure 5.18a, which shows a schematic of the dependence of $S(q^*)^{-1}$ on $T^{-1}$, as well as that of the domain spacing $d = 2\pi/q^*$, which changes slope within the disordered phase. The transition temperature to mean field behaviour is indicated as $T_{MF}$ and the extrapolated mean field spinodal temperature is denoted $T_{S,MF}$. Figure 5.18b shows an example of experimental SANS data showing the nonlinear behaviour of the inverse structure factor for a poly(ethylene-propylene)-*b*-poly(ethylethylene) (PEP-PEE, with a partially deuterated PEE block) diblock copolymer with a volume fraction of PEP, $f = 0.55$.

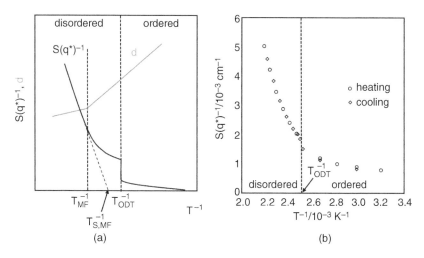

**Figure 5.18** (a) Schematic of behaviour of the inverse structure factor $S(q^*)^{-1}$ and domain spacing $d$ versus $1/T$ near the order-disorder transition temperature ($T_{ODT}$) of a diblock copolymer melt. *Source:* Redrawn from Sakamoto and Hashimoto [67]. (b) Experimental SANS data for a poly(ethylene-propylene)-$b$-poly(ethylethylene) PEP-PEE (with a partially deuterated PEE block) diblock copolymer melt with a volume fraction of PEP, $f = 0.55$, plotted as $S(q^*)^{-1}$ versus $1/T$. *Source:* Redrawn from Bates et al. [68].

# 5.9    KINETIC SANS USING LABELLED MIXTURES

Time-resolved small-angle neutron scattering (TR-SANS) has been successfully employed to measure the exchange kinetics of chains from block copolymer micelles, using mixtures of solutions of diblock copolymer micelles comprising perdeuterated or fully hydrogenous chains [69, 70]. The separate solutions were mixed at $t = 0$ in the kinetic studies (Figure 5.19). The average contrast of the poly(ethylene-propylene)-$b$-poly(ethylene oxide) (PEP-$b$-PEO) diblock copolymer was matched to that of the water/DMF solvent (selective for PEO) so that after complete chain exchange and mixing of deuterated and nondeuterated copolymer molecules, the scattering vanishes, permitting kinetic analysis of the decay of the intensity profiles. The experiments showed a logarithmic decay of the relaxation function $R(t)$ (obtained from the integrated intensity [71]), contrary to the expectations of the existing theory that predicted an activated process with single exponential (first-order) kinetics [69, 70]. This method was later used to investigate chain exchange in ordered (BCC) block copolymer micelle lattices [72], and exchange of block copolymer chains with different architectures [73], among other systems.

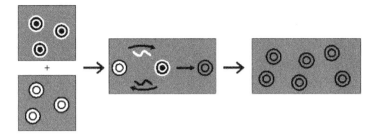

**Figure 5.19** Schematic of concept of time-resolved SANS measurements of chain exchange in diblock copolymer micelles. White and black indicate fully deuterated or hydrogenous chains, respectively. The average contrast is matched to that of the solvent (grey). *Source:* From Lund et al. [69]. © 2006, American Chemical Society.

Among other examples, TR-SANS was used to investigate the transition from lamellar phase to microemulsion in a ternary oil/water/surfactant system using a stopped-flow device [74]. The time resolution was approximately six seconds. In another study, TR-SANS with stopped-flow mixing was used to probe the transition from disc-like micelles to unilamellar vesicles in a mixed zwitterionic/anionic surfactant system [75]. The time resolution was better than 0.1 second (faster time resolution is not limited by the SANS flux, but is due to the mixing time for the sample volume 0.2–0.5 ml required for SANS measurements [76]). Figure 5.20 shows SANS data along with a schematic of the process. This system had previously been extensively studied by TR-SAXS [78, 79].

Other examples of TR-SANS have been reviewed [76].

## 5.10   ULTRA-SMALL-ANGLE SANS (USANS)

Among other applications [10], USANS is used to investigate the structure of porous materials (discussed in Section 5.11), phase separated polymer blends and microemulsions [80–82], aggregation (cluster formation) of colloids [83–85] hydrogel formation of biomolecules including peptides [86], protein network structures [87], [88], and the structure of emulsions [89–91]. Figure 5.21 shows an example of the combination of USANS and SANS data for an attractive colloidal glass formed by block copolymer micelles, along with a model fit, which shows a fractal like scaling at low $q$ (discussed in Section 1.6.5).

Figure 5.22 shows an example of combined SANS and USANS covering a very wide $q$ range of measurements (on several instruments, indicated)

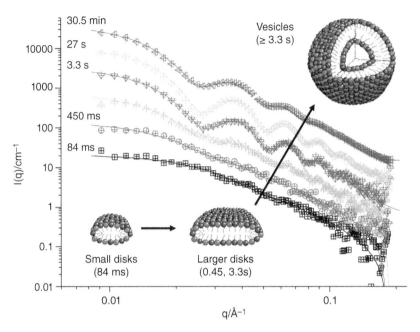

**Figure 5.20** TR-SANS study of transition from disc-like micelles to vesicles in a mixed surfactant system, tetradecyl dimethyl amine oxide 27.5 mM / 22.5 mM lithium perfluorooctane sulfonate in $D_2O$. *Sources:* From Gradzielski [76]; Hollamby [77].

on a sedimentary rock sample, which exhibits a surface fractal structure giving a power law intensity decay over many decades in intensity and $q$.

The analysis of USANS data often provides a fractal dimension for aggregate structures and/or a correlation length of an aggregate or pore structure. Several expressions for the structure/form factor of fractal aggregate structures are available; examples are given in Section 1.6.5. USANS and USAXS studies of different types of materials have been reviewed [10, 85]. The following section gives examples of measurements on other porous structures.

## 5.11   SANS AND USANS (AND SAXS AND USAXS) STUDIES ON POROUS STRUCTURES

SANS and USANS are particularly powerful methods to probe the structure of porous materials due to the ability to vary the contrast using mixtures of labelled solvents within the porous material [10]. SAXS and USAXS are also

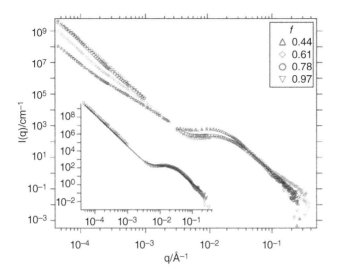

**Figure 5.21**   USANS data at low $q$ combined with SANS data at high $q$ (note gap in between the data sets) for polystyrene-$b$-poly(acrylic acid) PS-$b$-PAA block copolymer micelles forming a colloidal glass, $f$ denotes the degree of hydrolysis of the PAA block. The inset shows an example fit (red line) to the data that highlights the low $q$ power law scaling $I(q) \sim q^{-d}$ (with $d$ in the range 2.8–3.7 for the samples studied) and the high $q$ SANS data fitted to a model of the form/structure factor of hard spheres. *Sources*: From Hollamby [77], Crichton and Bhatia [92].

used to investigate porous materials. These methods provide information on the fractal structure of many porous materials, the pore size distribution as well as the local pore structure. Figure 5.23 shows representative SANS data from a nanoporous carbon sample showing regimes where information on different length scales can be extracted.

The combination of USANS and SANS provides information on pore structure over multiple length scales, ranging from the pore network structure and/or pore size distribution probed by USANS (or SANS at low $q$) to the size, internal shape, and structure of the pores probed by SANS (Figure 5.23). Contrast variation using solvent mixtures can provide additional information on the pore structure. As discussed in Section 5.10, the low $q$ scattering may exhibit a power law behaviour due to a fractal structure. Porous materials investigated by SANS (with or without USANS) include coal and related minerals, porous glasses, filled polymers, packings of colloidal particles, xero- and aero-gels, powders, micro-porous and meso-porous inorganic materials and others [77, 95]. A table of SANS studies on these systems is available [95]. These structures have also widely been investigated using SAXS and USAXS [96–98].

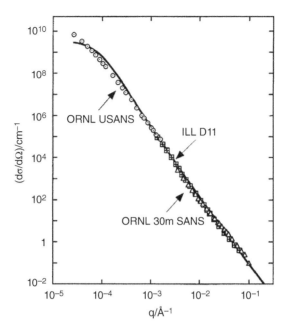

**Figure 5.22** Combination of SANS and USANS data for a sedimentary rock sample showing fractal structure (fractal dimension $D = 2.82$, correlation length $\zeta = 1.2\,\mu m$). The data are shown as symbols, the line is a fit to a model fractal structure factor. ORNL: Oak Ridge National Laboratory, ILL: Institut Laue-Langevin. *Source:* From Radlinski et al. [93]. © 1999, Amerian Physical Society.

The analysis of SAXS/SANS (USAXS/USANS) data from porous systems can provide the specific volume, the scattering density contrast, the invariant, the pore volume fraction and the strut volume fraction, and the pore and solid chord lengths (Eq. (2.20)) [99]. The latter can be interpreted, respectively, as the average pore size and average network strut size [99]. The specific volume is given by Eq. (2.19), while the invariant is given by Eq. (2.27) with $\Phi = \phi_p$, the volume fraction occupied by pores. SAXS and SANS have been combined to investigate pore size and shape in a good example of a study on a porous material, focussed on microporous carbon [100]. The Porod length and invariant were obtained, as well as the chord length distribution, among other quantities. A Porod plot enabled the specific surface area $S/m$ (surface area per unit mass) to be obtained from the Porod law, with a modification introduced by Ruland and coworkers [101] to allow for fluctuation scattering from the distribution of pore sizes:

$$\lim_{q\to\infty} I(q) = \frac{(2\pi)^4 P_m}{q^4} + \frac{B_{fl}}{q^2} \qquad (5.29)$$

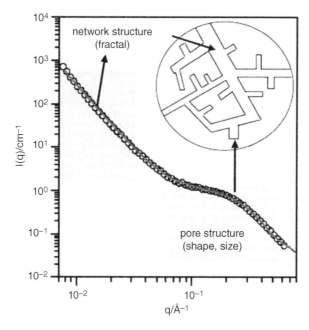

**Figure 5.23** SANS data from a nanoporous carbon sample with schematic of structure on different length scales. *Sources:* Original data from [94], adapted from Hollamby [77].

Here, $B_{fl}$ is the intensity for the fluctuation component, and the Porod constant, which provides the specific surface area is given by [100]

$$P_m = \frac{(\Delta\rho)^2}{(2\pi)^3} \frac{S}{m}$$
(5.30)

Figure 5.24 shows a modified Porod plot, the red line enabled the fluctuation scattering term to be obtained from the gradient and the Porod constant $P_m$ from the intercept, providing the specific surface area.

In another example, USAXS and USANS have been combined in a study on a mixed cellulose ester membrane [99]. GISANS and GISAXS have also been used to investigate porous structures in thin films, these techniques being discussed in Chapter 6.

## 5.12 SANS ON MAGNETIC MATERIALS

### 5.12.1 Unpolarized Neutron Beam

An important application of SANS is in the investigation of magnetic materials. For the case of spherical magnetic nanoparticles in solution (e.g. in a

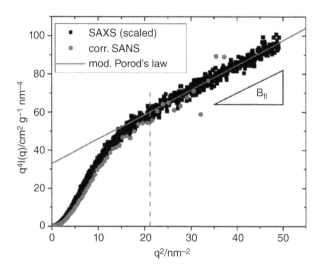

**Figure 5.24** Modified Porod plot for a microporous carbon sample, showing combined SAXS and SANS data. The red line is a fit to a modified Porod law, described in the text. *Source:* From Jafta et al. [100]. Licensed under CC BY 4.0.

ferrofluid) in the presence of a magnetic field $H$ (Figure 5.25) in an unpolarized neutron beam, in addition to the nuclear scattering given by Eq. (5.6) there is an additional magnetic contribution to the scattering cross-section [102–104]:

$$\frac{d\sigma_{mag}}{d\Omega} = [V\Delta\rho_{mag}F_s(q,R)]^2\sin^2\alpha \tag{5.31}$$

Here $F_s(q,R)$ is the form factor of a sphere given in Table 2.2, $\alpha$ is the angle between $\mathbf{q}$ and the magnetization contrast $\Delta\mathbf{M}$, and $\Delta\rho_{mag}$ is the magnetic analogue of the nuclear SLD, which is given by [105]

$$\Delta\rho_{mag} = \frac{1}{2}\frac{\gamma e\mu_0}{2\pi\hbar}\Delta\mathbf{M} \tag{5.32}$$

Here $\gamma = 1.913$ denotes the neutron magnetic moment expressed in units of the nuclear magneton [105–107], $e$ is the elementary charge, $\mu_0$ is the vacuum permeability ($1.2566 \times 10^{-6}$ H m$^{-1}$), and $\Delta\mathbf{M} = \mathbf{M}_p - \mathbf{M}$ is the contrast between the magnetization of the particle and that of the surrounding medium.

Conventional SANS with an unpolarized beam may be used to investigate vortex (flux line) lattices in type II superconductors among others. Figure 5.26 shows examples of SANS patterns from such structures showing the Bragg peaks in samples aligned in magnetic fields. As well as information on the structure of the lattices (Figure 5.26), the technique provides

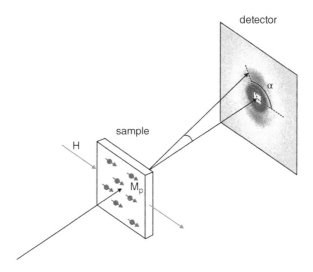

**Figure 5.25** SANS from aligned magnetic nanoparticles with magnetization $\mathbf{M}_p$ in the direction of the applied magnetic field $H$. The angle $\alpha$ parameterizes the azimuthal intensity variation.

quantitative measurements of penetration depth, lattice coherence length and electron (Cooper) pairing. SANS studies on vortex lattice structures have been reviewed [108–110].

## 5.12.2 Polarized Neutron Beams

SANS with polarized neutron beams can be used to investigate magnetic domain structures in alloys, and the structure of nanocrystalline magnetic materials and nanocomposites.

A flipper may be used to switch the polarization of neutrons in a beam. A polarizer is placed before the flipper before the sample in the beam path (cf. Figure 5.25), and the neutron beam passes through an analyser after the sample. The polarizer and analyser are ferromagnetic single crystals or $^3$He spin filters and the spin flipper is based on an electromagnetic coil, of which there are several designs.

There are three possible magnetic SANS experiments using (i) unpolarised neutrons, (ii) polarized incident beam but no polarization analysis, called SANSPOL, (iii) spin-resolved SANS, called POLARIS measurements. Depending on whether the initial spin is + or − and whether it is flipped or not, there are four possible cross-sections $\frac{d\sigma^{++}}{d\Omega}$, $\frac{d\sigma^{--}}{d\Omega}$ (non spin-flipped) and $\frac{d\sigma^{+-}}{d\Omega}$, $\frac{d\sigma^{-+}}{d\Omega}$ (spin flipped).

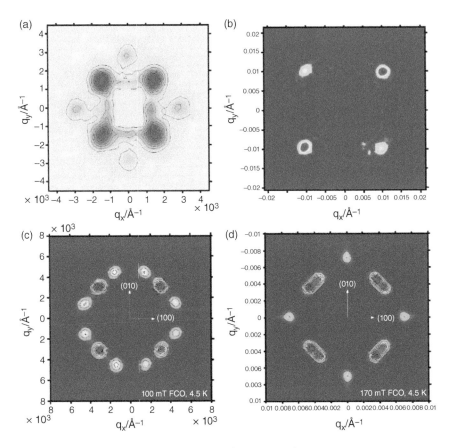

**Figure 5.26** Examples of SANS patterns from vortex lattices (VL) in type II super-conductors (all with $B$ parallel to the $c$ axis of the lattice). (a) square VL in the unconventional superconductor $Sr_2RuO_4$ at 50 mK, B = 0.025 T, (b) square VL in the high temperature superconductor $La_{2-x}Sr_xCuO_4$ at 2 K, B = 1.2 T, (c and d) low- and high-field rhombic VL in $YNi_2B_2C$ at 4.5 K, at (c) 0.1 T and (d) 0.17 T. *Source:* Adapted from Dewhurst and Cubitt [108].

The full expressions for the scattering cross-sections in the three sets of experiments depend on the magnetization vector field and the orientation of field with respect to the incident beam. The equations are rather complex and further details can be found elsewhere [104, 107].

Figure 5.27 shows an example of SANSPOL data obtained for a Co ferrofluid system, with the neutron spin directions indicated with respect to a horizontal magnetic field. This type of measurement corresponds to a contrast variation experiment using the neutron spin, which can easily be flipped (leaving the nonmagnetic contrast unchanged). The patterns

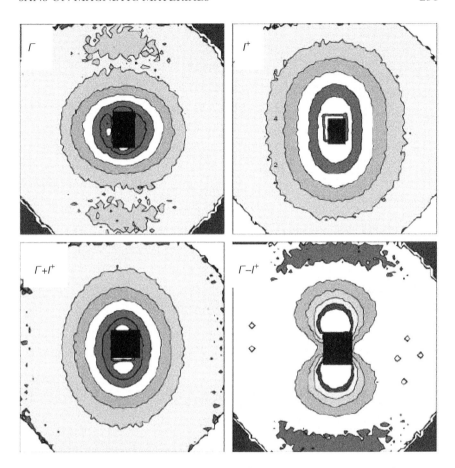

**Figure 5.27** SANSPOL data for a Co ferrofluid with the neutron spin directions indicated (top two patterns) with respect to a horizontal 1.1 T magnetic field. Bottom patterns: The arithmetic mean $[I^- + I^+]/2$ corresponds to the 2D pattern observed for nonpolarised neutrons. The difference $I^- - I^+$ yields nuclear-magnetic cross term scattering. The $q$ scale extends to $q = 3\,\mathrm{nm}^{-1}$ approximately. *Source:* From Wiedenmann [103].

in Figure 5.27 show very distinct anisotropy, depending on the neutron spin direction. This data yielded detailed information on the SLD at the surfactant-coated interface of the magnetic particles and the suspending fluid, as well as their magnetization properties [103]. The sum and difference scattering patterns shown in Figure 5.27 show a $\sin^2\alpha$ dependence from Eq. (5.31). The sum scattering corresponds to that of an unpolarized neutron scattering pattern, and the difference is related to a magnetic-nuclear scattering cross term [103]. Rather than a uniform sphere form factor, a core-shell form factor (Table 1.2) was used to describe the scattering from the surfactant-coated Co nanoparticles.

## 5.13   SPIN-ECHO SANS (SESANS)

SESANS is a technique that exploits the Larmor precession of neutrons in a magnetic field and analyses the depolarization of a neutron beam at small angles (Figure 5.28). Larmor precession refers to the rotation of the magnetic moment of the neutron about a magnetic field (in an electromagnetic coil in a spin echo neutron instrument). The angular frequency of precession is known as the Larmor frequency.

The depolarization is measured as a function of the spin echo length, which can be related to a correlation function. The measured depolarization is given by [112, 114]

$$\frac{P(z)}{P_0} = \exp[G(z) - G(0)] \tag{5.33}$$

Here $G(z)$ is a correlation function defined in Eq. (5.35) below, $P(z)$ is the polarization, $P_0$ is the incident beam polarization, and $z$ is the spin-echo length, which is defined as [111]

$$z = \frac{m\gamma}{2\pi h} BL\lambda^2 cot\, \eta_0 \tag{5.34}$$

Here $m$ is the neutron magnetic moment, $\gamma$ is the neutron gyromagnetic ratio (Section 5.12), $B$ is the magnetic flux density (field strength), $L$ is the length of the precession coils along the neutron beam direction, $\lambda$ is the neutron wavelength and $\eta_0$ is the magnetic field inclination angle (Figure 5.28).

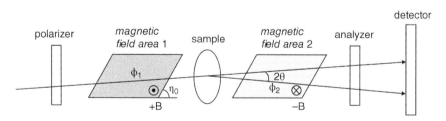

**Figure 5.28**   Schematic of SESANS setup [111–113]. The neutron precesses with tilt angle $\phi_1$ through the first inclined magnetic field before being scattered by the sample and precesses with tilt angle $\phi_2$ through a second inclined magnetic field (with opposite direction to the first precession field). Diffraction of the polarized neutron beam by a small angle $2\theta$ will result in $\Delta\phi = \phi_1 - \phi_2 \neq 0$, measured as depolarization. *Sources*: Based on Major et al. [111]; van Heijkamp et al. [113].

The correlation function is a projection of the scattering density autocorrelation function: [112, 114]

$$G(z) = \int_z^\infty \frac{\gamma(r)dr}{\sqrt{r^2 - z^2}} \tag{5.35}$$

Here $\gamma(r)$ is the scattering density autocorrelation function from Eq. (1.16).

SESANS is a technique not yet in widespread use, and to date initial studies have investigated, for example, aggregation and particle size in dairy products and colloids. The method is available on only a few instruments internationally, such as LARMOR at ISIS and at the Reactor Institute Delft. Further details on SESANS can be found elsewhere [114–117].

## 5.14  COMPLEMENTARY SAXS AND SANS

Combining SAXS with SANS gives an additional contrast from the SAXS experiment to add to whichever sample contrasts have been measured by SANS (some examples have been provided in this chapter). Complementary measurements have been used, to give a few examples, in the analysis of lipid bilayer structures [118, 119], block copolymer micelles [29, 120], nanoparticles [121], protein complex structures in solution [122–124], nanorods [125], nanoparticle-protein complexes [126], precipitates in metals and alloys [77, 127], and surfactant micelles [128–130], In the latter case, combined SAXS and SANS can provide information counterion condensation [129] and on the origin of peaks in the SAS data, whether from form or structure factor effects [130].

## REFERENCES

1. Otomo, T. (2011). Small-Angle Neutron Scattering. In: *Neutrons in Soft Matter* (eds. T. Imae, T. Kanaya, M. Furusaka and N. Torikai). Wiley Online Library: Wiley.
2. Narayanan, T. (2014). Small-Angle Scattering. In: *Structure from Diffraction Methods* (eds. D.W. Bruce, D. O'Hare and R.I. Walton). Wiley Online Library: Wiley.
3. Roe, R.-J. (2000). *Methods of X-ray and Neutron Scattering in Polymer Science*. New York: Oxford University Press.
4. Lindner, P. (2002). Scattering Experiments. In: *Neutrons, X-rays and Light Scattering Methods Applied to Soft Condensed Matter* (eds. P. Lindner and T. Zemb). Amsterdam: Elsevier.
5. Dianoux, A.-J. and Lander, G. (eds.) (2003). *Neutron Data Booklet*. Grenoble, France: Institut Laue-Langevin.
6. Narayanan, T. (2008). Synchrotron small-angle X-ray scattering. In: *Soft Matter Characterization* (eds. R. Borsali and R. Pecora), 899–952. Berlin: Springer-Verlag.

7. https://neutrons.ornl.gov/sites/default/files/GPSANS.pdf 2020.
8. Glinka, C.J., Barker, J.G., Hammouda, B. et al. (1998). The 30 m small-angle neutron scattering instruments at the National Institute of Standards and Technology. *Journal of Applied Crystallography* 31: 430–445.
9. Dewhurst, C.D. (2008). D33 - a third small-angle neutron scattering instrument at the Institut Laue Langevin. *Measurement Science and Technology* 19: 8.
10. Melnichenko, Y.B. (2016). *Small-Angle Scattering from Confined and Interfacial Fluids.* Berlin: Springer-Verlag.
11. Frielinghaus, H. (2007). Nanostructures investigated by small-angle neutron scattering. In: *Neutron Scattering: Lectures of the JCNS Laboratory Course held at Forschungszentrum Juelich and the research reactor FRM II of TU Munich* (eds. T. Brückel, G. Heger, D. Richter, et al.). Jülich, Germany: Forschungszentrum Jülich.
12. Svergun, D.I., Koch, M.H.J., Timmins, P.A., and May, R.P. (2013). *Small Angle X-ray and Neutron Scattering from Solutions of Biological Macromolecules.* Oxford: Oxford University Press.
13. Jeffries, C.M., Graewert, M.A., Svergun, D.I., and Blanchet, C.E. (2015). Limiting radiation damage for high-brilliance biological solution scattering: practical experience at the EMBL P12 beamline PETRAIII. *Journal of Synchrotron Radiation* 22: 273–279.
14. NIST Center for Neutron Research. (2020). Neutron scattering lengths and cross sections. https://www.ncnr.nist.gov/resources/n-lengths.
15. Pedersen, J.S., Posselt, D., and Mortensen, K. (1990). *Journal of Applied Crystallography* 23: 321.
16. Pedersen, J.S. (2002). Instrumentation for small-angle X-ray and neutron scattering. In: *Neutrons, X-rays and Light Scattering Methods Applied to Soft Condensed Matter* (eds. P. Lindner and T. Zemb). Amsterdam: Elsevier.
17. Berk, N.F. and Hardman-Rhyne, K.A. (1988). Analysis of SAS data dominated by multiple-scattering. *Journal of Applied Crystallography* 21: 645–651.
18. Pauw, B.R. (2013). Everything SAXS: small-angle scattering pattern collection and correction. *Journal of Physics: Condensed Matter* 25: 24.
19. Zaccai, N.R., Sandlin, C.W., Hoopes, J.T. et al. (2016). Deuterium labeling together with contrast variation small-angle neutron scattering suggests how Skp captures and releases unfolded outer membrane proteins. In: *Isotope Labeling of Biomolecules - Applications*, Methods in Enzymology, vol. 566 (ed. Z. Kelman), 159–210. San Diego: Elsevier Academic Press Inc.
20. Williams, C.E. (1991). Contrast variation in x-ray and neutron scattering. In: *Neutron, X-ray and Light Scattering* (eds. P. Lindner and T. Zemb). Amsterdam: Elsevier.
21. Stuhrmann, H.B. (1974). Neutron small-angle scattering of biological macromolecules in solution. *Journal of Applied Crystallography* 7: 173–178.
22. Bates, F.S., Dierker, S.B., and Wignall, G.D. (1986). Phase-behavior of amorphous binary-mixtures of perdeuterated and normal 1,4-polybutadienes. *Macromolecules* 19: 1938–1945.
23. Bates, F.S. and Wignall, G.D. (1986). Isotope-induced quantum-phase transitions in the liquid-state. *Physical Review Letters* 57: 1429–1432.
24. Bates, F.S., Keith, H.D., and McWhan, D.B. (1987). Isotope effect on the melting temperature of nonpolar polymers. *Macromolecules* 20: 3065–3070.
25. Bates, F.S. (1988). Small-angle neutron-scattering from amorphous polymers. *Journal of Applied Crystallography* 21: 681–691.
26. Wignall, G.D. and Bates, F.S. (1988). Applications and limitations of deuterium labeling methods to neutron-scattering studies of polymers. *Makromolekulare Chemie-Macromolecular Symposia* 15: 105–122.
27. Rhee, J. and Crist, B. (1993). Isotope and microstructure interactions in blends of random copolymers. *Journal of Chemical Physics* 98: 4174–4182.

28. Matsushita, Y., Mori, K., Mogi, Y. et al. (1990). Chain conformation of a block polymer in a microphase-separated structure. *Macromolecules* 23: 4317–4321.

29. Pedersen, J.S., Hamley, I.W., Ryu, C.Y., and Lodge, T.P. (2000). Contrast variation small-angle neutron scattering study of the structure of block copolymer micelles in a slightly selective solvent at semidilute concentrations. *Macromolecules* 33: 542–550.

30. Pedersen, J.S. and Gerstenberg, M.C. (1996). Scattering form factor of block copolymer micelles. *Macromolecules* 29: 1363–1365.

31. Pedersen, J.S. and Svaneborg, C. (2002). Scattering from block copolymer micelles. *Current Opinion in Colloid and Interface Science* 7: 158–166.

32. Kříž, J., Pleštil, J., Tuzar, Z. et al. (1998). Three layer micelles in an ABC block copolymer: NMR, SANS and LS study of poly(2-ethylhexyl acrylate)-block-poly(methyl methacrylate). *Macromolecules* 31: 41–51.

33. Hamley, I.W. (1998). *The Physics of Block Copolymers*. Oxford: Oxford University Press.

34. Hamley, I.W. and Castelletto, V. (2004). Small-angle scattering of block copolymers in the melt, solution and crystal states. *Progress in Polymer Science* 29: 909–948.

35. Hamley, I.W. (ed.) (2004). *Developments in Block Copolymer Science and Technology*. Chichester: Wiley.

36. Hamley, I.W. (2005). *Block Copolymers in Solution*. Chichester: Wiley.

37. Chen, S.H., Chang, S.L., Strey, R. et al. (1991). Structural evolution of bicontinuous microemulsions. *Journal of Physical Chemistry* 95: 7427–7432.

38. Knoll, W., Schmidt, G., Ibel, K., and Sackmann, E. (1985). Small-angle neutron-scattering study of lateral phase-separation in dimyristoylphosphatidylcholine cholesterol mixed membranes. *Biochemistry* 24: 5240–5246.

39. Knoll, W., Schmidt, G., and Ibel, K. (1985). The inverse contrast variation in small-angle neutron-scattering - a sensitive technique for the evaluation of lipid phase-diagrams. *Journal of Applied Crystallography* 18: 65–70.

40. Knoll, W., Haas, J., Stuhrmann, H.B. et al. (1981). Small-angle neutron-scattering of aqueous dispersions of lipids and lipid mixtures - a contrast variation study. *Journal of Applied Crystallography* 14: 191–202.

41. Whitten, A.E., Jacques, D.A., Hammouda, B. et al. (2007). The structure of the KinA-Sda complex suggests an allosteric mechanism of histidine kinase inhibition. *Journal of Molecular Biology* 368: 407–420.

42. Ibel, K. and Stuhrmann, H.B. (1975). Comparison of neutron and x-ray-scattering of dilute myoglobin solutions. *Journal of Molecular Biology* 93: 255–265.

43. Moore, P.B. (1982). Small-angle scattering techniques for the study of biological macromolecules and macromolecular aggregates. In: *Methods of Experimental Physics*, vol. 20 (eds. G. Ehrenstein and H. Lecar), 337–390. New York: Academic Press.

44. Whitten, A.E., Cai, S.Z., and Trewhella, J. (2008). MULCh: modules for the analysis of small-angle neutron contrast variation data from biomolecular assemblies. *Journal of Applied Crystallography* 41: 222–226.

45. Heller, W.T. (2010). Small-angle neutron scattering and contrast variation: a powerful combination for studying biological structures. *Acta Crystallographica, Section D: Biological Crystallography* 66: 1213–1217.

46. Engelman, D.M. and Moore, P.B. (1975). Determination of quaternary structure by small-angle neutron-scattering. *Annual Review of Biophysics and Bioengineering* 4: 219–241.

47. Sarachan, K.L., Curtis, J.E., and Krueger, S. (2013). Small-angle scattering contrast calculator for protein and nucleic acid complexes in solution. *Journal of Applied Crystallography* 46: 1889–1893.

48. Maric, S., Lind, T.K., Raida, M.R. et al. (2019). Time-resolved small-angle neutron scattering as a probe for the dynamics of lipid exchange between human lipoproteins and naturally derived membranes. *Scientific Reports* 9: 14.

49. Ober, R., Cotton, J.P., Farnoux, B., and Higgins, J.S. (1974). Calculation of neutron-diffraction pattern by polymer-chains in bulk state. *Macromolecules* 7: 634–641.
50. Williams, C.E., Nierlich, M., Cotton, J.P. et al. (1979). Polyelectrolyte solutions - intrachain and interchain correlations observed by SANS. *Journal of Polymer Science Part C-Polymer Letters* 17: 379–384.
51. Akcasu, A.Z., Summerfield, G.C., Han, C.C. et al. (1980). Measurement of single chain neutron-scattering in concentrated polymer solutions. *Journal of Polymer Science Part B-Polymer Physics* 18: 863–869.
52. Wignall, G.D., Hendricks, R.W., Koehler, W.C. et al. (1981). Measurements of single chain form-factors by small-angle neutron-scattering from polystyrene blends containing high-concentrations of labeled molecules. *Polymer* 22: 886–889.
53. Benoit, H. and Benmouna, M. (1984). Scattering from a polymer-solution at an arbitrary concentration. *Polymer* 25: 1059–1067.
54. Richardson, R.M., Allman, J.M., and McIntyre, G. (1990). Neutron scattering from mixtures of isotopically labelled molecules. A new method for determining the orientational distribution function in liquid crystals. *Liquid Crystals* 7: 701–719.
55. Higgins, J.S. and Benoît, H.C. (1994). *Polymers and Neutron Scattering*. Oxford: Oxford University Press.
56. Date, R.W., Hamley, I.W., Luckhurst, G.R. et al. (1992). Orientational ordering in liquid crystals: isotope labelling neutron diffraction experiments. *Molecular Physics* 76: 951–977.
57. Hamley, I.W., Garnett, S., Luckhurst, G.R. et al. (1996). Orientational ordering in the nematic phase of a thermotropic liquid crystal: a small angle neutron scattering study. *Journal of Chemical Physics* 104: 10046–10054.
58. Hammouda, B. (1993). SANS from homogeneous polymer mixtures - a unified overview. *Advances in Polymer Science* 106: 87–133.
59. Hammouda, B. (2010). SANS from polymers-review of the recent literature. *Polymer Reviews* 50: 14–39.
60. Wang, Y., Wang, W., Hong, J.Y. et al. (2020). Quantitative examination of a fundamental assumption in small-angle neutron scattering studies of deformed polymer melts. *Polymer* 204: 122698.
61. Pedersen, J.S. and Schurtenberger, P. (1999). Static properties of polystyrene in semidilute solutions: a comparison of Monte Carlo simulation and small-angle neutron scattering results. *Europhysics Letters* 45: 666–672.
62. Janssen, S., Schwahn, D., and Springer, T. (1992). Mean-Field Ising crossover and the critical exponens $\gamma$, $\nu$ and $\eta$ for a polymer blend - d-PB/PS studied by small-angle neutron-scattering. *Physical Review Letters* 68: 3180–3183.
63. Schwahn, D., Mortensen, K., and Yee-Madeira, H. (1987). Mean-Field and Ising critical-behavior of a polymer blend. *Physical Review Letters* 58: 1544–1546.
64. Schwahn, D., Mortensen, K., Springer, T. et al. (1987). Investigation of the phase-diagram and critical fluctuations of the system polyvinylmethylether and d-polystyrene with neutron small-angle scattering. *Journal of Chemical Physics* 87: 6078–6087.
65. Bates, F.S., Rosedale, J.H., Fredrickson, G.H., and Glinka, C.J. (1988). Fluctuation-induced first order phase transition of an isotropic system to a periodic state. *Physical Review Letters* 61: 2229–2232.
66. Bates, F.S., Rosedale, J.H., Stepanek, P. et al. (1990). Static and dynamic crossover in a critical polymer mixture. *Physical Review Letters* 65: 1893–1896.
67. Sakamoto, N. and Hashimoto, T. (1995). Order-disorder transition of low molecular weight polystyrene-block-polyisoprene. 1. SAXS analysis of two characteristic temperatures. *Macromolecules* 28: 6825–6834.
68. Bates, F.S., Rosedale, J.H., and Fredrickson, G.H. (1990). Fluctuation effects in a symmetric diblock copolymer near the order-disorder transition. *Journal of Chemical Physics* 92: 6255–6270.

69. Lund, R., Willner, L., Richter, D., and Dormidontova, E.E. (2006). Equilibrium chain exchange kinetics of diblock copolymer micelles: tuning and logarithmic relaxation. *Macromolecules* 39: 4566–4575.

70. Lund, R., Willner, L., Stellbrink, J. et al. (2006). Logarithmic chain-exchange kinetics of diblock copolymer micelles. *Physical Review Letters* 96: 068302.

71. Lund, R., Willner, L., and Richter, D. (2013). Kinetics of block copolymer micelles studied by small-angle scattering methods. In: *Controlled Polymerization and Polymeric Structures: Flow Microreactor Polymerization, Micelles Kinetics, Polypeptide Ordering, Light Emitting Nanostructures*, Advances in Polymer Science, vol. 259 (eds. A. Abe, K.S. Lee, L. Leibler and S. Kobayashi), 51–158. Berlin: Springer-Verlag Berlin.

72. Choi, S.H., Bates, F.S., and Lodge, T.P. (2011). Molecular exchange in ordered diblock copolymer micelles. *Macromolecules* 44: 3594–3604.

73. Lu, J., Bates, F.S., and Lodge, T.P. (2015). Remarkable effect of molecular architecture on chain exchange in triblock copolymer micelles. *Macromolecules* 48: 2667–2676.

74. Tabor, R.F., Eastoe, J., and Grillo, I. (2009). Time-resolved small-angle neutron scattering as a lamellar phase evolves into a microemulsion. *Soft Matter* 5: 2125–2129.

75. Bressel, K., Muthig, M., Prevost, S. et al. (2010). Mesodynamics: watching vesicle formation in situ by small-angle neutron scattering. *Colloid and Polymer Science* 288: 827–840.

76. Gradzielski, M. (2012). Dynamics of self-assembled systems studied by neutron scattering: Current state and perspectives. *European Physical Journal Special Topics* 213: 267–290.

77. Hollamby, M.J. (2013). Practical applications of small-angle neutron scattering. *Physical Chemistry Chemical Physics* 15: 10566–10579.

78. Gummel, J., Sztucki, M., Narayanan, T., and Gradzielski, M. (2011). Concentration dependent pathways in spontaneous self-assembly of unilamellar vesicles. *Soft Matter* 7: 5731–5738.

79. Narayanan, T., Gummel, J., and Gradzielski, M. (2014). Probing the self-assembly of unilamellar vesicles using time-resolved SAXS. In: , Advances in Planar Lipid Bilayers and Liposomes, vol. 20 (eds. A. Iglic and C.V. Kulkarni), 171–196. San Diego: Elsevier Academic Press Inc.

80. Agamalian, M., Alamo, R.G., Kim, M.H. et al. (1999). Phase behavior of blends of linear and branched polyethylenes on micron length scales via ultra-small-angle neutron scattering. *Macromolecules* 32: 3093–3096.

81. Lee, J.H., Ruegg, M.L., Balsara, N.P. et al. (2003). Phase behavior of highly immiscible polymer blends stabilized by a balanced block copolymer surfactant. *Macromolecules* 36: 6537–6548.

82. Wanakule, N.S., Nedoma, A.J., Robertson, M.L. et al. (2008). Characterization of micron-sized periodic structures in multicomponent polymer blends by ultra-small-angle neutron scattering and optical microscopy. *Macromolecules* 41: 471–477.

83. Harada, T., Matsuoka, H., Yamamoto, T. et al. (2001). The structure of colloidal alloy crystals revealed by ultra-small-angle neutron scattering. *Colloids and Surfaces a-Physicochemical and Engineering Aspects* 190: 17–24.

84. Bhatia, S., Barker, J., and Mourchid, A. (2003). Scattering of disklike particle suspensions: evidence for repulsive interactions and large length scale structure from static light scattering and ultra-small-angle neutron scattering. *Langmuir* 19: 532–535.

85. Bhatia, S.R. (2005). Ultra-small-angle scattering studies of complex fluids. *Current Opinion in Colloid and Interface Science* 9: 404–411.

86. Pakstis, L.M., Ozbas, B., Hales, K.D. et al. (2004). Effect of chemistry and morphology on the biofunctionality of self-assembling diblock copolypeptide hydrogels. *Biomacromolecules* 5: 312–318.

87. Weigandt, K.M., Pozzo, D.C., and Porcar, L. (2009). Structure of high density fibrin networks probed with neutron scattering and rheology. *Soft Matter* 5: 4321–4330.

88. Whittaker, J.L., Balu, R., Knott, R. et al. (2018). Structural evolution of photocrosslinked silk fibroin and silk fibroin-based hybrid hydrogels: a small angle and ultra-small

angle scattering investigation. *International Journal of Biological Macromolecules* 114: 998–1007.

89. Innerlohinger, J., Villa, M., Baron, M., and Glatter, O. (2006). Ultra-small-angle neutron scattering from dense micrometre-sized colloidal systems: data evaluation and comparison with static light scattering. *Journal of Applied Crystallography* 39: 202–208.

90. Zank, J., Reynolds, P.A., Jackson, A.J. et al. (2006). Aggregation in a high internal phase emulsion observed by SANS and USANS. *Physica B: Condensed Matter* 385: 776–779.

91. Whitby, C.P., Djerdjev, A.M., Beattie, J.K., and Warr, G.G. (2007). In situ determination of the size and polydispersity of concentrated emulsions. *Langmuir* 23: 1694–1700.

92. Crichton, M. and Bhatia, S. (2003). Structure and intermicellar interactions in block polyelectrolyte assemblies. *Journal of Applied Crystallography* 36: 652–655.

93. Radlinski, A.P., Radlinska, E.Z., Agamalian, M. et al. (1999). Fractal geometry of rocks. *Physical Review Letters* 82: 3078–3081.

94. Tsao, C.S., Li, M., Zhang, Y. et al. (2010). Probing the room temperature spatial distribution of hydrogen in nanoporous carbon by use of small-angle neutron scattering. *Journal of Physical Chemistry C* 114: 19895–19900.

95. Ford, K.M., Konzman, B.G., and Rubinson, J.F. (2011). A more informative approach for characterization of polymer monolithic phases: small angle neutron scattering/ultrasmall angle neutron scattering. *Analytical Chemistry* 83: 9201–9205.

96. Schaefer, D.W. and Keefer, K.D. (1986). Structure of random porous materials - silica aerogel. *Physical Review Letters* 56: 2199–2202.

97. Egger, C.C., du Fresne, C., Raman, V.I. et al. (2008). Characterization of highly porous polymeric materials with pore diameters larger than 100 mn by mercury porosimetry and x-ray scattering methods. *Langmuir* 24: 5877–5887.

98. Yokoyama, H. (2013). Small angle x-ray scattering studies of nanocellular and nanoporous structures. *Polymer Journal* 45: 3–9.

99. Hu, N.P., Borkar, N., Kohls, D., and Schaefer, D.W. (2011). Characterization of porous materials using combined small-angle x-ray and neutron scattering techniques. *Journal of Membrane Science* 379: 138–145.

100. Jafta, C.J., Petzold, A., Risse, S. et al. (2017). Correlating pore size and shape to local disorder in microporous carbon: a combined small angle neutron and x-ray scattering study. *Carbon* 123: 440–447.

101. Perret, R. and Ruland, W. (1968). X-ray small-angle scattering of non-graphitizable carbons. *Journal of Applied Crystallography* 1: 308–313.

102. Kohlbrecher, J., Wiedenmann, A., and Wollenberger, H. (1997). Magnetic coupling between the different phases in nanocrystalline Fe-Si-B studied by small angle neutron scattering. *Zeitschrift für Physik B: Condensed Matter* 104: 1–4.

103. Wiedenmann, A. (2001). Small-angle neutron scattering investigations of magnetic nanostructures and interfaces using polarized neutrons. *Physica B* 297: 226–233.

104. Michels, A. (2014). Magnetic small-angle neutron scattering of bulk ferromagnets. *Journal of Physics: Condensed Matter* 26: 41.

105. Squires, G.L. (2012). *Introduction to the Theory of Thermal Neutron Scattering*. Cambridge: Cambridge University Press.

106. Lovesey, S.W. (1984). *Theory of Neutron Scattering from Condensed Matter*, Polarization Effects and Magnetic Scattering, vol. 2. Oxford: Oxford University Press.

107. Muhlbauer, S., Honecker, D., Perigo, E.A. et al. (2019). Magnetic small-angle neutron scattering. *Reviews of Modern Physics* 91: 75.

108. Dewhurst, C.D. and Cubitt, R. (2006). Small-angle neutron scattering (SANS) studies of the vortex lattice in type II superconductors. *Physica B: Condensed Matter* 385: 176–179.

109. Eskildsen, M.R., Forgan, E.M., and Kawano-Furukawa, H. (2011). Vortex structures, penetration depth and pairing in iron-based superconductors studied by small-angle neutron scattering. *Reports on Progress in Physics* 74: 13.

110. Eskildsen, M.R. (2011). Vortex lattices in type-II superconductors studied by small-angle neutron scattering. *Frontiers of Physics* 6: 398–409.

111. Major, J., Dosch, H., Felcher, G.P. et al. (2003). Combining of neutron spin echo and reflectivity: a new technique for probing surface and interface order. *Physica B: Condensed Matter* 336: 8–15.

112. Tromp, R.H. and Bouwman, W.G. (2007). A novel application of neutron scattering on dairy products. *Food Hydrocolloids* 21: 154–158.

113. van Heijkamp, L.F., de Schepper, I.M., Strobl, M. et al. (2010). Milk gelation studied with small angle neutron scattering techniques and Monte Carlo simulations. *Journal of Physical Chemistry. A* 114: 2412–2426.

114. Krouglov, T., de Schepper, I.M., Bouwman, W.G., and Rekveldt, M.T. (2003). Real-space interpretation of spin-echo small-angle neutron scattering. *Journal of Applied Crystallography* 36: 117–124.

115. Rekveldt, M.T. (1996). Novel SANS instrument using neutron spin echo. *Nuclear Instruments & Methods in Physics Research Section B-Beam Interactions with Materials and Atoms* 114: 366–370.

116. Rekveldt, M.T., Plomp, J., Bouwman, W.G. et al. (2005). Spin-echo small angle neutron scattering in Delft. *Review of Scientific Instruments* 76.

117. Andersson, R., van Heijkamp, L.F., de Schepper, I.M., and Bouwman, W.G. (2008). Analysis of spin-echo small-angle neutron scattering measurements. *Journal of Applied Crystallography* 41: 868–885.

118. Kucerka, N., Nagle, J.F., Sachs, J.N. et al. (2008). Lipid bilayer structure determined by the simultaneous analysis of neutron and x-ray scattering data. *Biophysical Journal* 95: 2356–2367.

119. Kucerka, N., Nieh, M.P., and Katsaras, J. (2011). Fluid phase lipid areas and bilayer thicknesses of commonly used phosphatidylcholines as a function of temperature. *Biochimica et Biophysica Acta - Biomembranes* 1808: 2761–2771.

120. Lund, R., Willner, L., Richter, D. et al. (2013). Kinetic pathway of the cylinder-to-sphere transition in block copolymer micelles observed in situ by time-resolved neutron and synchrotron scattering. *ACS Macro Letters* 2: 1082–1087.

121. Schindler, T., Schmiele, M., Schmutzler, T. et al. (2015). A combined SAXS/SANS study for the in situ characterization of ligand shells on small nanoparticles: the case of ZnO. *Langmuir* 31: 10130–10136.

122. Svergun, D.I. and Nierhaus, K.H. (2000). A map of protein-rRNA distribution in the 70 S *Escherichia coli* ribosome. *The Journal of Biological Chemistry* 275: 14432–14439.

123. Niemann, H.H., Petoukhov, M.V., Hartlein, M. et al. (2008). X-ray and neutron small-angle scattering analysis of the complex formed by the met receptor and the Listeria monocytogenes invasion protein InlB. *Journal of Molecular Biology* 377: 489–500.

124. Mirandela, G.D., Tamburrino, G., Ivanovic, M.T. et al. (2018). Merging in-solution x-ray and neutron scattering data allows fine structural analysis of membrane-protein detergent complexes. *Journal of Physical Chemistry Letters* 9: 3910–3914.

125. Gomez-Grana, S., Hubert, F., Testard, F. et al. (2012). Surfactant (Bi)layers on gold nanorods. *Langmuir* 28: 1453–1459.

126. Spinozzi, F., Ceccone, G., Moretti, P. et al. (2017). Structural and thermodynamic properties of nanoparticle protein complexes: a combined SAXS and SANS study. *Langmuir* 33: 2248–2256.

127. Ohnuma, M., Suzuki, J., Ohtsuka, S. et al. (2009). A new method for the quantitative analysis of the scale and composition of nanosized oxide in 9Cr-ODS steel. *Acta Materialia* 57: 5571–5581.

128. Aswal, V.K., Goyal, P.S., De, S. et al. (2000). Small-angle x-ray scattering from micellar solutions of Gemini surfactants. *Chemical Physics Letters* 329: 336–340.

129. Aswal, V.K., Goyal, P.S., Amenitsch, H., and Bernstorff, S. (2004). Counterion condensation in ionic micelles as studied by a combined use of SANS and SAXS. *Pramana-Journal of Physics* 63: 333–338.
130. Zemb, T. and Diat, O. (2010). What can we learn from combined SAXS and SANS measurements of the same sample containing surfactants? In: *XIV International Conference on Small-Angle Scattering*, Journal of Physics Conference Series., vol. 247 (ed. G. Ungar). Bristol: IOP Publishing Ltd.

# 6

# Grazing-Incidence Small-Angle Scattering

## 6.1 INTRODUCTION

This chapter presents the basic principles of grazing incidence small-angle x-ray scattering (GISAXS) and grazing-incidence small-angle neutron scattering (GISANS), along with some examples of GIWAXS. It also covers x-ray and neutron reflectivity. These techniques provide a wealth of information on the ordering of samples at interfaces including thin films, such as the scattering length density (SLD) profile (at Å resolution) normal to the surface in the case of reflectivity measurements, and quantitative measurements of roughness from specular reflectivity measurements. Off-specular reflection and GISAS experiments also provide unique information on ordering including that in the plane of the surface, i.e. on lateral structures present in soft materials, as well as inorganic systems such as nanoparticle films, magnetic nanostructures and others. This chapter does not cover 'surface science' aspects of grazing incidence diffraction (GID), i.e. on measurements of metal and metal oxide surfaces under ultrahigh vacuum (UHV) conditions.

The GISAXS measurements discussed here can be performed using lab or synchrotron instruments for transmission SAXS with modification of the scattering geometry. The GISANS measurements can similarly be performed on modified transmission SANS beamlines as well as custom beamlines. The reflectivity experiments are usually performed on dedicated x-ray or neutron beamlines. In addition, commercial x-ray reflectometers are available, or this geometry can be provided as part of a powder diffraction instrument.

*Small-Angle Scattering: Theory, Instrumentation, Data and Applications*,
First Edition. Ian W. Hamley.
© 2021 John Wiley & Sons Ltd. Published 2021 by John Wiley & Sons Ltd.

This chapter provides a concise summary of the key equations used in the interpretation of GISAXS and GISANS data but does not give a full account of the theory that is available elsewhere [1].

This chapter is organized as follows. Section 6.2 introduces basic quantities including definition of angles, refractive index and SLD. The different types of specular and off-specular scans that are performed in a reflectivity experiment to obtain particular structural information are discussed in Section 6.3. The distorted wave Born approximation (DWBA), which is essential to properly analyse GISAXS data to provide detailed structural information is introduced in Section 6.4. The analysis of grazing-incidence small-angle scattering (GISAS) data is discussed in Section 6.5. Informative and representative examples of experimental GISAS data from selected systems including polymer films and nanoparticles deposited at surfaces, along with a few examples of in situ and kinetic measurements are presented in Section 6.6. Finally, Section 6.7 gives examples of experimental GIWAXS/GIXD (grazing-incidence x-ray diffraction) data.

## 6.2   BASIC QUANTITIES: DEFINITION OF ANGLES, REFRACTIVE INDEX, AND SCATTERING LENGTH DENSITY

GISAXS and GISANS are becoming popular techniques to perform structural studies on the ordering at the surface of thin films of samples of inorganic and soft materials. Good reviews on this subject are available [1–5]. The x-ray or neutron beam is incident at a grazing angle of incidence and the reflected beam is detected (Figure 6.1).

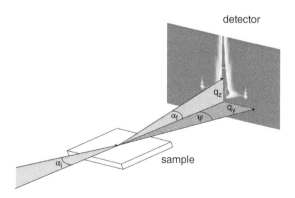

**Figure 6.1**   GISAS (grazing-incidence small-angle scattering) geometry, with conventional notation for angles and wavevector components.

The x-rays or neutrons are incident at a glancing angle $\alpha_i$ (up to a few degrees, as in all small-angle scattering experiments) and are scattered to a vertical angle $\alpha_f$. This plane of reflection is termed the *specular plane*. The components of wavevector **q** are [1, 4, 5]:

$$q_x = \frac{2\pi}{\lambda}(\cos\psi\cos\alpha_f - \cos\alpha_i),$$

$$q_y = \frac{2\pi}{\lambda}(\sin\psi\cos\alpha_f),$$

$$q_z = \frac{2\pi}{\lambda}(\sin\alpha_f + \sin\alpha_i), \qquad (6.1)$$

Scattering along the $z$ direction is termed specular reflection. The technique of specular reflectivity is a defined technique, which is used to determine the structure (electron or SLD profile) perpendicular to the surface of a film. This technique can be performed on specialist instruments and a fuller discussion is available elsewhere [6–11]. Off-specular reflection arises in the case that the sample has lateral structure or roughness.

The reflection of x-rays and neutrons arises because of the refractive index difference at an interface, the critical angle is very small (as in grazing incidence) when the refractive index $n < 1$. The refractive index for x-rays and neutrons is given by [2, 12–15]

$$n = 1 - \delta + i\beta \qquad (6.2)$$

Here, $\delta$ is related to the SLD of coherent scattering, and $\beta$ is proportional to the absorption cross-section. For x-rays, these coefficients are given by [2, 4, 12, 15, 16]

$$\delta = \frac{r_e\rho\lambda^2}{2\pi}, \beta = \frac{\mu\lambda}{4\pi} \qquad (6.3)$$

Here, $\lambda$ is the wavelength of the x-rays, $r_e$ is the Thomson or classical scattering length of an electron (Eq. (4.6)), $\rho$ is the electron density of the medium, and $\mu$ is the linear absorption coefficient (Eq. (2.16)).

For neutrons there are analogous equations [4, 12, 13, 16]:

$$\delta = \frac{\rho_a b\lambda^2}{2\pi}, \beta = \frac{\rho_a\sigma_a\lambda}{4\pi} \qquad (6.4)$$

Here, $\rho_a$ is the number density of atoms with scattering length $b$ and $\sigma_a$ is the absorption cross-section. The product $\rho_a b$ is the neutron SLD. Typically, for either x-rays or neutrons, $\delta \sim 10^{-6} - 10^{-5}$ and $\beta \sim 10^{-8}$; therefore, $n$ is only slightly less than unity. Often the absorption term containing $\beta$ is neglected [8].

In a grazing incidence scattering experiment with x-rays or neutrons, the fact that $n < 1$ leads to total reflection below a critical angle, which is characteristic of the material. The critical angle (in radians) is given approximately by [14–16]:

$$\alpha_c = \sqrt{2\delta} = \lambda\sqrt{\rho/\pi} \tag{6.5}$$

Total reflection is observed for $\alpha_f < \alpha_c$. Since $\delta$ is small, $\alpha_c$ is typically ~milliradians. However, some x-rays or neutrons do penetrate into the surface even below $\alpha_c$, producing an evanescent wave. The intensity of the evanescent wave decays exponentially with depth [1, 12, 15]. Close to the critical angle, the penetration depth $\tau_{1/e}$ is given by [17]

$$\tau_{1/e} = \frac{2^{1/2}\lambda}{4\pi}[((\alpha_i^2 - \alpha_c^2)^2 + 4\beta^2)^{1/2} - (\alpha_i^2 - \alpha_c^2)]^{-1/2} \tag{6.6}$$

Figure 6.2 shows an example of a calculation using this equation for a silicon surface. Below the critical angle, $\tau_{1/e}$ tends to a constant value $\tau_{1/e} \approx \lambda/4\pi\alpha_c$ [17].

The evanescent wave is mainly exploited in spectroscopic measurements, but its presence modifies the absorption coefficient $\beta$ and it can act as a secondary source for interference between scattered waves, and therefore GISAXS can be performed below $\alpha_c$, although it is typically performed up to a few multiples of $\alpha_c$. Table 6.1 lists values of $\alpha_c$ for neutrons and x-rays for selected materials, along with neutron scattering length densities and electron densities. Tables (with some examples for polymers) are also provided in refs. [15, 16, 20, 21].

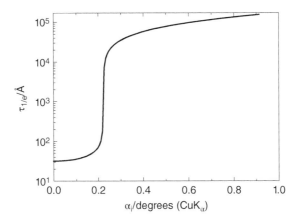

**Figure 6.2** Penetration depth of x-rays calculated for a Si surface using Eq. (6.6). Calculated using the calculator http://www-ssrl.slac.stanford.edu/materialscatter/gixs-calculator.xls, with allowance for linear attenuation coefficient.

**Table 6.1** Table of scattering length densities and critical angles for GISAXS and GISANS[a,b].

| Material | $\rho$ (x-ray)[c] $/10^{-6}$ Å$^{-2}$ | $\delta$ (x-ray) $/10^{-6}$ | $\alpha_c$ (x-ray) / ° | $\rho$ (neu) $/10^{-6}$ Å$^{-2}$ | $\delta$ (neu) $/10^{-6}$ | $\alpha_c$ (neu) / ° |
|---|---|---|---|---|---|---|
| Si | 20.06 | 7.59 | 0.223 | 2.073 | 0.783 | 0.072 |
| Au | 12.48 | 47.11 | 0.556 | 4.667 | 1.761 | 0.108 |
| $H_2O$ | 9.469 | 3.574 | 0.153 | −0.561 | −0.212 | – |
| $D_2O$ | 9.455 | 3.569 | 0.153 | 6.393 | 2.413 | 0.126 |
| Polystyrene | 9.699 | 3.661 | 0.155 | 1.426 | 0.538 | 0.059 |
| Deuterated polystyrene | 9.512 | 3.590 | 0.154 | 6.407 | 2.418 | 0.126 |
| Polyethylene | 8.926 | 3.369 | 0.149 | −0.329 | −0.124 | – |
| Deuterated polyethylene | 8.994 | 3.395 | 0.149 | 7.959 | 3.004 | 0.140 |
| Poly(methyl Methacrylate) | 10.84 | 4.092 | 0.164 | 1.059 | 0.400 | 0.051 |

[a]*Sources:* Based on data in refs. [15, 18, 19].
[b]Critical angles are calculated with $\lambda = 1.54$ Å.
[c]This quantity is $r_e\rho$ in Eq. (6.3).

## 6.3   CHARACTERISTIC SCANS

### 6.3.1   Specular Reflectivity ($z$ Scan)

Consider refraction and reflection at a single interface (Figure 6.3) within the $(x, z)$ plane in Figure 6.3. The case that $\alpha_f = \alpha_i$ is termed *specular reflection and scanning* through $\alpha_i$ $(= \alpha_f)$, which defines $q_z = 4\pi\sin(\alpha)/\lambda$ and measuring the reflectivity is termed a *specular reflectivity measurement.*

We consider henceforth the amplitudes of both the reflected and transmitted waves [2, 15, 22, 23], since these are required to calculate reflectivity

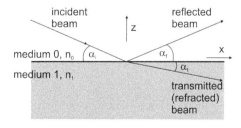

**Figure 6.3** Reflection and refraction at an interface between two media with refractive indices shown along with the definition of angles used in the text. Here $t$ denotes transmission (equivalent to refraction).

from a multilayer or when considering the DWBA (Section 6.4), but note that in the reflectivity experiment it is the reflected intensity that is measured.

Snell's law (recall elementary physics) relates the angle of refraction across an interface to the refractive indexes for the two media:

$$n_0 \cos \alpha_i = n_1 \cos \alpha_t \qquad (6.7)$$

This can be written as

$$n_0^2(1 - \sin^2 \alpha_0) = n_1^2(1 - \sin^2 \alpha_t) \qquad (6.8)$$

Since the critical angle is given by $n_0 \cos \alpha_c = n_1$ (in air, $n_0 = 1$),

$$n_1^2 \sin^2 \alpha_t = n_0^2 \sin^2 \alpha_i - n_0^2 \sin^2 \alpha_c \qquad (6.9)$$

This can be written in terms of wavenumbers as

$$q_t^2 = q^2 - q_c^2 \qquad (6.10)$$

where $q = 2k \sin \alpha_i$ (with incident $k = k_i = 2\pi n_0/\lambda$), with related expressions for the refracted wavevector and critical wavevectors.

The Fresnel coefficient for the reflected amplitude is [2, 8, 16, 19, 21, 23–25]

$$r(q) = \frac{q - q_t}{q + q_t} \qquad (6.11)$$

For the transmitted amplitude, the Fresnel coefficient is

$$t(q) = \frac{2q}{q + q_t} \qquad (6.12)$$

The Fresnel equation for the reflectivity (reflected intensity) is given by

$$R = \left| \frac{q - q_t}{q + q_t} \right|^2 = \left| \frac{q - \sqrt{q^2 - q_c^2}}{q + \sqrt{q^2 - q_c^2}} \right|^2 \qquad (6.13)$$

By expanding Eq. (6.13) in terms of $q/q_c$, it can be shown that the Fresnel intensity decays as $q^{-4}$ [20]:

$$R_F = \frac{1}{16} \left( \frac{q_c}{q} \right)^4 \qquad (6.14)$$

It can be useful to use the expression, from Eq. (6.5) [8, 20],

$$q_c = \sqrt{16\pi \rho_1} \qquad (6.15)$$

For a rough interface, the reflectivity is modified to [13, 16, 20, 27–29]

$$R = R(q)\exp(-\sigma^2 q^2) \tag{6.16}$$

where $\sigma$ is the roughness (expressed in terms of Gaussian halfwidth). In fact, the roughness of liquid interfaces has been investigated in considerable detail based on an analysis of capillary waves (thermally excited waves at a liquid surface, the existence and properties of which depend on surface tension). This topic is beyond the scope of the present chapter. More information can be found in texts [25, 26] and many chapters and reviews that cover the subject in detail (see, for example, refs. [24, 30–32]). These contributions and others also cover the early literature on the important subject of reflectivity studies of amphiphilic molecules at liquid interfaces.

Figure 6.4 shows the Fresnel reflectivity calculated for a surface with no roughness, along with an example to illustrate the effect of roughness (with $\sigma = 10\,\text{Å}$) and a reflectivity profile for a monolayer slab showing Kiessig fringes. These are characteristic of the reflectivity profile of layered systems. The spacing of Kiessig fringes is inversely proportional to the thickness of the layer, $d = 2\pi/(\Delta q)$ [20].

These equations can be generalized to consider multilayer stacks, allowing for the reflection and transmission at each interface. The equations involve use of recursion relations for the reflection and transmission at a given interface, depending on the same processes at other interfaces that the beam

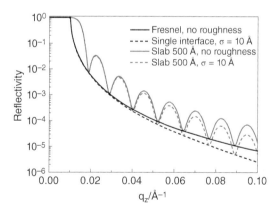

**Figure 6.4** Examples of reflectivity profiles for single interface showing reflectivity for an interface of a Si with SLD $\rho = 2.07 \times 10^{-6}\,\text{Å}^{-2}$ (Table 6.1), with and without surface roughness $\sigma = 10\,\text{Å}$, and for a monolayer slab of thickness $500\,\text{Å}$ without surface roughness or $\sigma = 10\,\text{Å}$ for both interfaces, with a slab neutron SLD $\rho = 4 \times 10^{-6}\,\text{Å}^{-2}$. Calculated using the NIST slab reflectivity calculator (available at the webpage shown in Table 2.2).

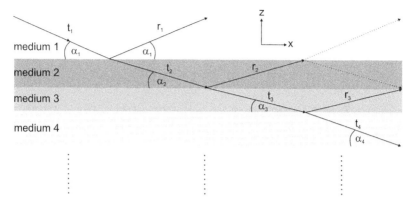

**Figure 6.5** Reflectivity from a multilayer system can be analysed using a recursion formula. The Fresnel amplitudes for reflected and transmitted waves are indicated by $r_n$ and $t_n$, respectively. Selected multiple reflection and refraction processes are indicated by dotted lines.

has passed through (Figure 6.5). The process can also be represented by a matrix formalism, the so-called Parratt method [26, 33]. The expressions are equivalent to those derived in the analysis of the optics of multilayer films [22, 23, 27, 34, 35]. Multiple reflection and transmission processes can be considered. Further information on this is available elsewhere [1, 8, 13, 15, 16, 20, 25, 26, 33]. Various software is available to perform the full dynamical theory calculations for multilayer or slab systems (see http://gisaxs.com/index.php/Software for a list), and via web interfaces (see, for example, http://smallangle.org/content/software#Model-Fitting).

As an alternative to the full recursive/matrix formalism within dynamical theory, an approximate expression that relates the reflectivity to the gradient of the scattering density is available using kinematical theory [8, 20, 36]:

$$\frac{R(q)}{R_F(q)} = \left| \frac{1}{\rho_{sub}} \int_{-\infty}^{\infty} \frac{d\rho(z)}{dz} \exp(-iq'z)dz \right|^2 \tag{6.17}$$

Here $R_F(q)$ is the Fresnel reflectivity for the substrate (Eq. (6.13)), $\rho_{sub}$ is the scattering density of the substrate and $q' = \sqrt{q^2 - q_c^2}$, where $q_c$ is the critical $q$ value for the substrate, is an approximate refraction-corrected $q$ value.

The approximations used in deriving this equation from the full dynamical theory are described elsewhere [8]. The approximation breaks down in particular at low angles as shown by the example in Figure 6.6, but is usually quite accurate at a wavenumber that is 2–3 multiples of $q_c$ or above.

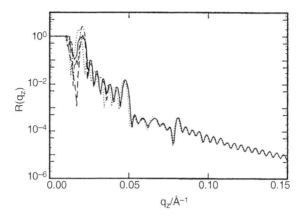

**Figure 6.6** Reflectivity profiles calculated using the full dynamical theory (solid line) and kinematical theory approximation (Eq. (6.17)) for a multilayer system, based on measurements for polystyrene-*b*-poly(methylmethacrylate) block copolymer film. *Source:* From Hamley and Pedersen [8]. © 1994, International Union of Crystallography.

Periodic multilayers structures, as formed by block copolymers in thin films or multilayer magnetic materials for example, will produce interference (quasi-Bragg) peaks. Figure 6.7 shows an example of x-ray and neutron reflectivity data measured from a lamellar triblock copolymer film. The x-ray contrast between the layers is extremely small; therefore, the x-ray reflectivity profile only contains Kiessig fringes due to the overall film thickness.

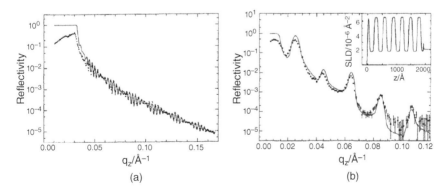

**Figure 6.7** Reflectivity profiles measured from (a) x-ray reflectivity and (b) neutron reflectivity on a P2VP-*d*PS-P2VP (P2VP: poly(2-vinylpyridine), *d*PS: deuterated polystyrene) triblock copolymer multilayer film. The open symbols are the measured data and the lines are calculations using the Parratt matrix formulation, discussed in the text. *Source:* From de Jeu et al. [37]. © 1993, EDP Sciences.

In this particular case, the Kiessig fringe intensity is modulated by the presence of islands at the surface of the block copolymer film (i.e. there are two film thickness, 1935 and 2229 Å) [37]. The neutron reflectivity data from the same block copolymer film on the other hand contains pseudo Bragg-peaks from the layer periodicity, with SLD shown in the inset of the neutron reflectivity profile in Figure 6.7b. In addition to lamellar block copolymer films, this multilayer type of SLD and reflectivity profile is observed for multilayer magnetic materials among other examples.

## 6.3.2   Other Scans

A scan along $q_y$ reveals the Yoneda (or Yoneda/Vineyard) peaks at an exit angle (Figure 6.1) $\alpha_f = \alpha_c$. This peak results from enhanced dynamical diffuse scattering, as discussed in more detail elsewhere [26]. Figure 6.8 shows an example of a GISAXS pattern with a horizontal cut shown, along with the corresponding intensity profile along this section, which shows the Yoneda peaks either side of the vertical specular reflection line.

Yoneda peaks can also be observed in detector scans, which are scans of the angle of the detector with respect to the horizontal (Figure 6.9a). They are observed in these off-specular scans at the critical angle, i.e. when $\beta = \alpha_c$.

Rocking scans, also known as transverse scans, give information on surface roughness through fitting of the width and shape of the specular peak. A rocking scan can be achieved by tilting of the sample as shown in Figure 6.9b. In a rocking scan $\alpha_i + \alpha_f$ is fixed. Figure 6.10 shows examples

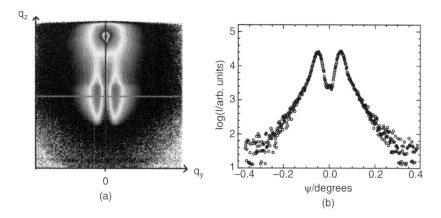

**Figure 6.8**   (a) GISAXS pattern (from a polystyrene-*b*-polyisoprene diblock copolymer film) along with (b) horizontal cut showing Yoneda peaks either side of the minimum at $q_y = 0$ (specular reflection). *Source:* From Müller-Buschbaum [3]. © 2003, Springer Nature.

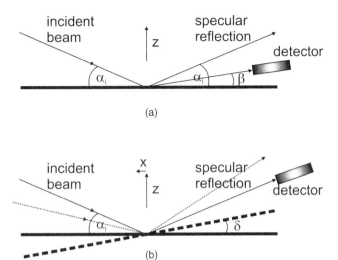

**Figure 6.9** (a) Detector scan of angle $\beta$, (b) Sample rocking curve, the sample is rocked by small angles $\delta$ (in both directions) around the specular condition. The result is a scan along $q_x$ (the normal to the surface moves essentially along $x$ for small rotation angles).

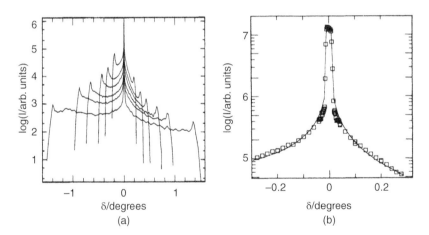

**Figure 6.10** Rocking curves (from x-ray reflectivity) (a) Polystyrene-*b*-poly(methyl methacrylate) diblock copolymer film [20], with Yoneda peaks evident. *Source:* From Russell [20]. © 1990, Elsevier. (b) Polished pyrex glass surface (symbols) with fit to a DWBA model for a surface with self-affine roughness [28]. Note the narrow range of $\delta$ for this plot of the central specular peak. *Source:* From Sinha et al. [28]. © 1988 American Physical Society.

of x-ray reflectivity rocking curves measured for a diblock copolymer film and for a polished Pyrex glass surface (in a narrower angle range close to the specular peak), along with a fit to the model for self-affine surfaces in the DWBA (see Section 6.4). Beyond the central specular peak, the Yoneda peaks are present in the wings in the profiles in Figure 6.10a. These can also be modelled in the DWBA [25, 28]. The asymmetry in the intensity of the peaks is simply due to the fact that a larger area of the sample is illuminated for negative $\delta$.

## 6.4 THE DISTORTED WAVE BORN APPROXIMATION

The DWBA goes beyond the Born approximation in considering multiple scattering events leading to the distortion of the (x-ray or neutron) wave as it passes from one medium to another. In the Born approximation, the incident wave is scattered independently at each point. In the DWBA, the scattering is calculated based on the Born approximation, but treating the effect of multiple scattering and reflection events in the medium in a perturbative fashion. The DWBA was developed for rough and patterned surfaces in several key papers [28, 38, 39], and has been reviewed in detail in the context of GISAXS [1, 25]. Following these papers, in the DWBA theory the scattered intensity is represented as a differential scattering cross-section.

For a rough surface the differential cross-section for diffuse scattering, close to the specular condition is given by [5, 20, 28, 39, 40]

$$\frac{d\sigma}{d\Omega} = L_x L_y \frac{\pi^2}{\lambda^4}(1 - n^2)^2|t(q_i)|^2|t(q_f)|^2 S(q') \qquad (6.18)$$

Here $L_x L_y$ is the footprint of the beam on the same (length $L_x$ and $L_y$), $n$ is the refractive index of the medium, $t(q)$ denote Fresnel transmission coefficients (Eq. (6.12)), and $S(q')$ is the structure factor, which is related to a Fourier transform of the height–height correlation function (the full equations are provided elsewhere [1, 20, 28, 39, 40]), with $q'$ indicating the wavenumber in the medium (of the rough surface material).

For patterned surfaces, i.e. those with lateral structures, for example nanoparticles deposited on a substrate, several geometries of which are shown in Figure 6.11 or block copolymer films (examples of GISAS data from which are discussed in Section 6.6), the differential scattering cross-section is factored into form and structure factors. As in the case of bulk transmission small-angle scattering as described in Section 1.5, various expressions can be used, including the local monodisperse approximation or the decoupling approximation. For example, in the local monodisperse

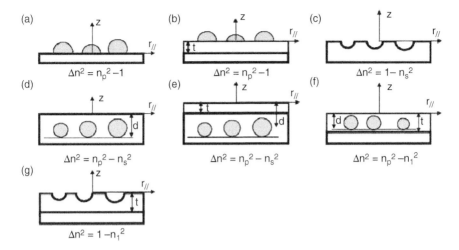

**Figure 6.11** Different arrangements of nanoparticles on surfaces along with indicated refractive indexes $n_p$: particle, $n_s$: substrate, $n_l$: layer and $n = 1$ (vacuum). (a) Nanoparticles/islands supported on a substrate, (b) nanoparticles on a thin layer of thickness $t$, (c) holes in a surface, (d) inclusions in a substrate, (e) inclusions in a substrate below a layer, (f) inclusions in a layer, (g) holes in a layer. *Source:* From Renaud et al. [1]. © 2009, Elsevier.

approximation, the differential cross-section can be written as [3, 41]

$$\frac{d\sigma}{d\Omega}(\mathbf{q}) = cP(\mathbf{q})S(\mathbf{q}) \tag{6.19}$$

where $c$ is a term that depends on the incident wavevector and the refractive indexes of the surface components (see Figure 6.11) and the form factor is $P(\mathbf{q}) = |F(\mathbf{q})|^2$ (with appropriate averaging over the size distribution). The decoupling approximation (Section 1.5) may be used as an alternative factoring of the form and structure factor. It can be noted that Eq. (6.19) is expressed in terms of anisotopic form and structure factors, since GISAS patterns are commonly aligned due to surface-induced ordering.

The form and structure factors are related to those for transmission small-angle scattering (SAS) discussed in Sections 1.5–1.7; however, in the DWBA the expressions are modified by the Fresnel coefficients. The equations are quite complex, and, for example, expressions are available for the structures shown in Figure 6.11 [1].

In one example, that of isolated nanoparticles on a substrate (shown in Figure 6.11a), neglecting the structure factor, the differential scattering cross section is given by [1, 5, 41]

$$\frac{d\sigma}{d\Omega}(\mathbf{q}) = \frac{k_0^4}{16\pi^2}|n_p^2 - 1|^2|\mathcal{F}(q_{xy}, k_{i,z}, k_{f,z})|^2 \tag{6.20}$$

Here, $k_0 = 2\pi/\lambda$, $n_p$ is the refractive index of the particle, $q_{xy}$ refers to the component of $\mathbf{q}$ in the $(x, y)$ plane and $q_z$ to that along $z$, $k_{i,z}$, $k_{f,z}$ are the wavevectors of incident and reflected waves, and

$$\mathcal{F}(q_{xy}, k_{i,z}, k_{f,z})$$
$$= F(q_{xy}, q_z) + r(\alpha_i)F(q_{xy}, p_z) + r(\alpha_f)F(q_{xy}, -p_z) + r(\alpha_i)r(\alpha_f)F(q_{xy}, -q_z) \tag{6.21}$$

where $p_z = (k_{i,z} + k_{f,z})$, $r(\alpha_i)$ and $r(\alpha_f)$ are Fresnel reflectivity coefficients (cf. Eq. (6.11)) and the functions $F(q_{xy}, \zeta)$ ($\zeta = q_z, p_z, -p_z, -q_z$) are anisotropic form factors, for example representing the hole as a cylinder (radius $R$ and length $L$; the following approximate formula can be used:

$$F(q_{xy}, \zeta) = 2\pi R^2 H \frac{J_1(q_{xy}R)}{q_{xy}R} \frac{\sin(\zeta L/2)}{\zeta L/2} \exp\left(-\frac{i\zeta L}{2}\right) \tag{6.22}$$

Here $J_1(x)$ denotes a first order Bessel function and this equation contains terms as for the transmission SAXS case considered in Section 1.7.3.

Approximate form factor formulae for other shapes are listed elsewhere [41], and expressions for structure factors used in DWBA calculations are also available [1]. Form factors for other surface geometries (for example those shown in Figure 6.11) analogous to Eq. (6.21) are presented elsewhere [1, 42].

In the Born approximation (BA), which becomes more reliable at high $q$, the expressions are simplified since the Fresnel reflectivity coefficients appearing in equations such as Eq. (6.21) become zero. Figure 6.12a shows calculated form factors for a cylinder of radius $R = 5$ nm and length $L = 5$ nm within the DWBA (Eq. (6.20)–(6.22)) and Born approximation (BA), and Figure 6.12b illustrates an example of the full DWBA calculation of a 2D GISAS pattern for aligned cylinders. In Figure 6.12a, there are quite significant differences in the position and depth of the form factor minima for $\alpha_i = \alpha_c$, and the BA approximation is not very reliable for GISAXS data in this region. It is a much better approximation away from $\alpha_c$ (e.g. $\alpha_i = 3\alpha_c$, shown in Figure 6.12a). However, as presented in Chapter 2 the Born approximation is perfectly adequate for transmission SAS since refraction and reflection effects are small in this case.

A GISAS pattern in the $(q_y, q_z)$ plane can be considered a distorted transmission SAS pattern. Powder diffraction rings become elliptical, and Bragg peaks may be elongated in the $q_z$ direction (due to the finite film thickness), becoming Bragg rods. On the other hand, scattering from long cylinder structures perpendicular to the substrate leads to Bragg sheets.

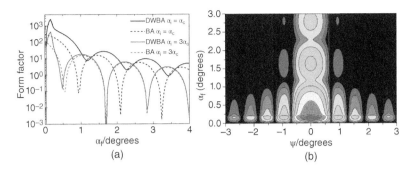

**Figure 6.12** (a) Calculation of the form factor (for specular reflectivity, i.e. $\psi = 0$) of cylinder-shaped islands on a surface (length = 5 nm, radius = 5 nm) at an angle of incidence $\alpha_i = \alpha_c = 0.2°$ and at $\alpha_i = 3\alpha_c$, comparing DWBA and BA calculations as indicated, (b) Simulated GISAXS pattern for the DWBA case with $\alpha_i = \alpha_c$ and the same cylinder dimensions as in part (a) and alignment along $z$ with Gaussian orientation distribution with width (standard deviation) $\sigma = 20°$. Both plots were generated using IsGISAXS [41].

Figure 6.13 summarizes GISAS patterns for different orientations of a lamellar structure, such as a block copolymer lamellar structure in a film at a solid surface. Lamellae that are parallel to the substrate (Figure 6.13a) give rise to Bragg peaks along $q_z$ arising from the one-dimensional periodicity perpendicular to the surface, and in addition there are diffuse stripes or Bragg sheets elongated along $q_y$. This type of structure is also observed for Langmuir-Blodgett multilayer films of amphiphiles such as lipids.

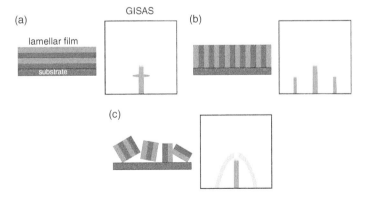

**Figure 6.13** Schematic of GISAS patterns from lamellar structures, (a) parallel to substrate, (b) perpendicular to substrate, (c) with mixed orientation and/or regions of disorder.

Perpendicular lamellae (Figure 6.13b) give rise to Bragg rods, elongated along $q_z$ and spaced along $q_y$ by $2\pi/D_{lam}$ where $D_{lam}$ is the lamellar period. GISAS patterns for films with a mixed orientation of lamellae (or with only partial lamellar ordering, along with regions of disorder) show a diffuse ring (ellipse) of scattering (Figure 6.13c).

Examples of experimental GISAXS data for block copolymer lamellar film structures are discussed in Section 6.6, which also shows examples of GISAS patterns from hexagonal and cubic lattices (structure factors) observed for different nanomaterials.

## 6.5  DATA ANALYSIS

Lists of software for GISAS and reflectivity data analysis and fitting are available in a database of other SAS software (see http://smallangle.org/content/software#Biomolecular-Fibre), and a very thorough list specific to GISAS and reflectivity data analysis is also valuable (see http://gisaxs.com/index .php/Software).

Data reduction, i.e. making corrections for background, detector response etc. is performed as for transmission SAS data described in Sections 2.4–2.6. These steps in data processing, along with translation to suitable $q$ or angle scales, may be performed using software listed on the web at http://gisaxs .com/index.php/Software. Software used for transmission SAS listed in Table 2.2 may sometimes be used for GISAS data, since it is often recorded on the same beamline with the same type of detector and GISAS-specific software is also available (Table 2.2)

Considering GISAXS modelling software, IsGISAXS can calculate form and structure factors within the DWBA [41]. NANOCELL Simulates 2D diffraction patterns from single crystals for GISAS, using the DWBA [44]. FitGISAXS [42] can fit GISAXS data; however, it requires commercial Igor-Pro software. HipGISAXS provides high-performance (massively parallel) software for simulating GISAXS data, based on embedding a collection of scatterers on a multilayered structure [45]. HipGISAXS works using generalized models for a series of custom and user-defined shaped objects, embedded in a multilayered structure. Scattering from objects embedded in layered films, without analytical expressions for their form factor, can be modelled, as well as those with closed form expressions for the form factor. BornAgain is recent Python-based software that extends the capabilities of IsGISAXS to include graded interfaces and with unrestricted numbers of layers and particles, and it allows for diffuse reflections from layer interfaces and particles with inner structure [46].

## 6.6   EXPERIMENTAL EXAMPLES OF GISAS DATA

The following gives a few examples of GISAS measurements for soft and hard nanostructured materials, excluding examples from 'surface science' studies performed under high vacuum, which are outside the scope of this text since these experiments require UHV sample environments, not available on many synchrotron or neutron beamlines where GISAS is done on the same instrument as transmission SAS, without specialist high-vacuum equipment. Further information on UHV GISAXS 'surface science' experiments (often at lower x-ray energy) is available elsewhere [1].

### 6.6.1   Block Copolymer Films

Block copolymers can undergo microphase separation into a range of ordered structures with typical 2–20 nm periodicity. Structures include lamellar, hexagonal, cubic, and bicontinuous morphologies, and these can align in different ways in thin films, depending on surface energies, annealing and solvent effects, etc. [47–49]. GISAS is a powerful means to study such structures since it offers the capability to probe in-plane and out-of-plane ordering, including depth profiling, which is not possible on native samples (i.e. in a nondestructive mode) with imaging techniques such as AFM or electron microscopy. In addition, GISAS provides ensemble averaged information on the ordering near interfaces, in contrast to the local imaging of microscopic methods. GISAS also enables in situ measurements, and this has been applied in a number of studies to examine the effects of solvent vapour annealing (used for morphology alignment control), as discussed in more detail in Section 6.6.3. GISAS measurements on block copolymer films have been reviewed [3–5]. Figure 6.14 shows an example of a GISAXS pattern from a block copolymer film comprising hexagonal-packed cylinders aligned perpendicular to the surface (silicon). The indexed Bragg reflections along $q_y$ are consistent with the in-plane hexagonal ordering shown in the AFM image in Figure 6.14, which reveals different orientations of the hexagonal lattice. (In this case, both $\sqrt{3}q^*$ and $4q^*$ reflections can be observed in the GISAXS pattern.) Bragg rods extending along the specular

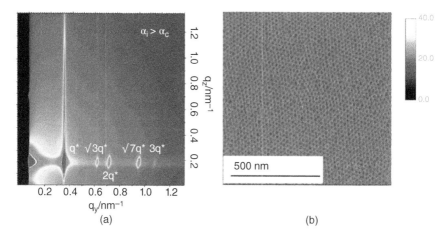

**Figure 6.14** (a) GISAXS pattern (measured at an incident angle $\alpha_i = 0.12°$, $\alpha_i > \alpha_c$, with $\alpha_c$ the critical angle of the copolymer film) with indexed reflections along $q_y$ ($q^* = 0.36\,\text{nm}^{-1}$ is the first-order peak position), and (b) phase contrast tapping mode AFM image for a thin film of a poly(cyclohexylethylene)-*b*-poly (ethylene)-*b*-poly(cyclohexylethylene) triblock copolymer forming a hexagonal-packed cylinder phase. *Source:* From Khanna et al. [50].

direction can be seen for the lower orders of reflection, resulting from the finite extent of the cylinders in the thin polymer film. In the GISAXS pattern in Figure 6.14a, the black area along $q_z$ around $q_y = 0$ shows a region blocked by beamstop to prevent saturation of the detector by the specular reflection of the beam. This masking is not necessary for GISANS experiments where the beam flux is much lower (Section 5.3).

## 6.6.2   Nanoparticles

GISAXS can be used to determine the structure of inorganic nanoparticles deposited on substrates. An example of a GISAXS pattern for a highly ordered surface structure of iron oxide nanocubes is shown in Figure 6.15, which contains many reflections from a body-centred tetragonal array, prepared by slow solvent evaporation from the Ge substrate in the presence of a magnetic field perpendicular to the wafer [51]. The crystalline stacking order perpendicular to the substrate leads to scattering from lattice planes with $l \neq 0$. Combining Bragg's law for a tetragonal lattice (cf. Eq. (1.55)) with cell parameters ($a$, $c$) and Snell's law (Eq. (6.10)) leads to an equation for

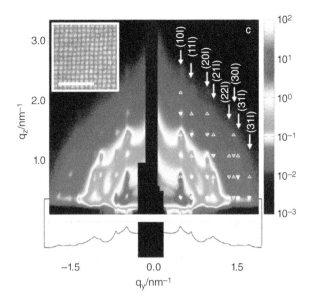

**Figure 6.15** GISAXS pattern for an ordered array of iron oxide nanocubes (SEM image shown in inset, scale bar = 100 nm) prepared by slow evaporation in the presence of a 30 mT magnetic field perpendicular to the substrate. The peaks are indexed according to a body-centred tetragonal structure, up and down diamonds indicate peaks observed with and without reflection from the substrate, respectively. The bottom profile is a horizontal slice taken at $q_z = 0.34$ nm$^{-1}$. The blacked-out area masks specular reflection. *Source:* From Disch et al. [51].

the position along $q_z$ of the observed reflections with Miller indices $(h, k, l)$:

$$q_z = k_{i,z} + \sqrt{k_{c,z}^2 + \left[ \sqrt{\left( 4\pi^2 \left( \frac{h^2 + k^2}{a^2} + \frac{l^2}{c^2} \right) \right)^2 - q_y^2} \pm \sqrt{k_{i,z}^2 - k_{c,z}^2} \right]^2} \tag{6.23}$$

Here $k_{i,z}$ is the $z$-component of the incident wavevector and $k_{c,z}$ is the corresponding wavevector at the critical angle. Similar equations apply to other structures (e.g. the equation for a lamellar structure is given in Eq. (6.24)).

An example showing a high degree of order in nanoparticle films is provided in a study using GISAXS to investigate the growth mode of gold nanoparticles templated by DNA assembly on a DNA-functionalized gold surface [43]. Figure 6.16 shows two orientations of a body-centred cubic

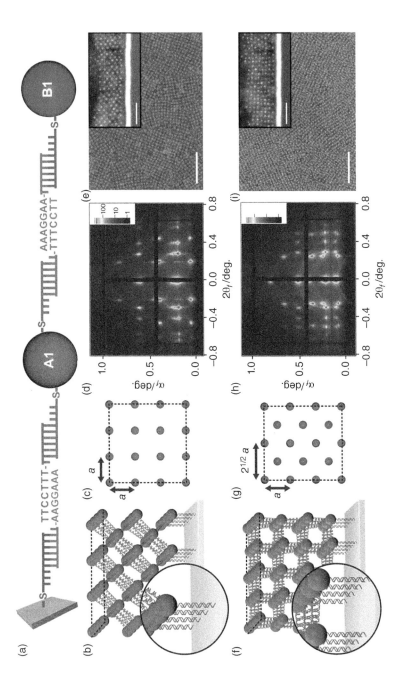

**Figure 6.16**  Step-by-step deposition of DNA-templated gold nanoparticle BCC arrays with two orientations shown. (a) Schematic of the system – DNA functionalized (attached via thiol) gold substrate (left) to which DNA-tethered gold nanoparticles (A1 and B1) are attached via successive immersion steps. (b–e) Schematic of (100) orientation of BCC structure (b, c) with corresponding GISAXS pattern in (d) and SEM image in (e), (f–i) Schematic of (110) orientation (f, g) with corresponding GISAXS pattern in (h) and SEM image in (i). Scale bars for the SEM top-down and cross-section views in (e) and (i) are 200 nm and 100 nm, respectively. Source: From Senesi et al.[12]

(BCC) lattice that could be obtained by successive deposition using DNA with different linkers.

The development of the thin film structure at each step of the deposition process was also examined using GISAXS [43].

Reviews in the literature discuss examples of many other GISAXS (and GIWAXS) studies of nanoparticle structures and growth on different substrates [1, 52].

## 6.6.3 Kinetic and In situ Studies

GISAXS has been used to probe phase transitions during controlled hydration of supported thin films (on a silicon substrate) of several lipids [53]. Figure 6.17 shows GISAXS data obtained from a film of the lipid phytantriol as hydration increases, along with a plot of relative humidity and a contour plot of the (radially integrated) intensity as a function of time. The system undergoes a transition from an crystalline/isotropic $L_2$ phase to lamellar ($L_\alpha$), then $Q_{II}^G$ gyroid and $Q_{II}^D$ diamond bicontinuous cubic phases as hydration increases.

Other studies that highlight the capabilities of time-resolved GISAXS include the work of Papadakis et al. on solvent annealing and drying effects on block copolymer lamellar films [54–57]. This group performed synchrotron GISAXS on films of symmetric polystyrene-$b$-polybutadiene diblock copolymers spin-coated onto a silicon substrate to produce an initial state with at least some lamellae parallel to the silicon substrate. The films were then annealed in vapours of toluene, cyclohexane, or ethyl acetate.

A time-resolved synchrotron GISAXS study of the structure of a symmetric diblock during solvent annealing and drying in ethyl acetate revealed an initial state of mixed lamellar alignment, followed by swelling and the eventual formation of a state of parallel lamellae [57]. These underwent de-swelling during drying. The process is shown schematically in Figure 6.18.

The parallel lamellae produce diffuse scattering rings (actually ellipses) termed diffuse Debye–Scherrer rings [57], as well as diffuse Bragg sheets spaced along the specular direction ($q_y = 0$) (cf. Figure 6.13). These features are shown in the series of GISAXS patterns measured during the swelling process (Figure 6.19).

The position of the diffuse Debye-Scherrer rings is given by [56–58]

$$q_z = k_{i,z} + \sqrt{k_{c,z}^2 + \left[ \sqrt{\left(\frac{2\pi m}{D_{\text{lam}}}\right)^2 - q_y^2} \pm \sqrt{k_{i,z}^2 - k_{c,z}^2} \right]^2} \qquad (6.24)$$

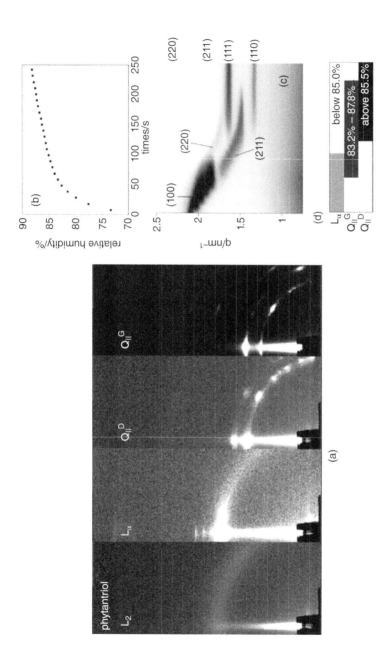

**Figure 6.17** (a) GISAXS patterns measured (L–R) as a function of increasing time in (b), along with (c) an intensity map of radially averaged data showing development of peaks in the ordered phases as shown schematically in (d). *Source:* From Rittman et al. [53]. © 2013, American Chemical Society.

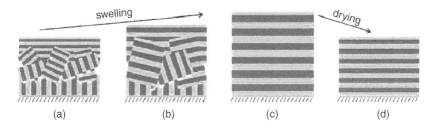

**Figure 6.18** Transitions observed via time-resolved GISAXS in a symmetric diblock copolymer film during solvent vapour annealing and subsequent drying. (a) Initial state at $t = 0$ s, (b) swollen film at $t = 828$ s, (c) swollen film at $t = 1350$ s, (d) dried film after t $= 7200$ s. *Source:* From Zhang et al. [57]. © 2014, American Chemical Society.

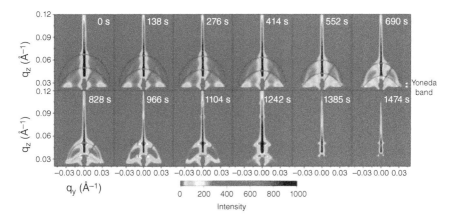

**Figure 6.19** Series of GISAXS patterns measured during solvent vapour annealing of a symmetric polystyrene-*b*-polybutadiene diblock copolymer in ethyl acetate [57]. The red rings fit the elliptical diffuse Debye-Scherrer rings (at intermediate times). The Yoneda band is indicated in the pattern taken at 690 s, and plus and minus branch first- and third-order diffuse Bragg sheets are labelled in the 966 s image. *Source:* Figure 2a from Zhang et al. [57].

Here, $k_{i,z}$ and $k_{c,z}$ are defined after Eq. (6.23), $D_{lam}$ is the lamellar spacing and $m$ is the order of diffraction. This group also presented theoretical expressions for the GISAS scattering cross-section in the DWBA for parallel and perpendicular lamellar structures [58].

The diffuse Bragg sheets along the specular direction fall into minus and plus branches, labelled M and P, respectively (corresponding to the $\pm$ in Eq. (6.24)), first and third orders of these are shown in the pattern at 966 s in Figure 6.19. During drying, further sharpening of the GISAXS pattern

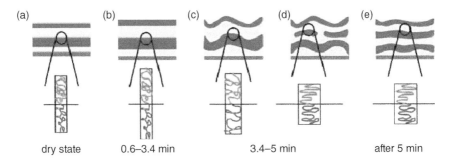

**Figure 6.20**  Processes occurring during toluene solvent annealing of a parallel lamellar symmetric diblock copolymer film, deduced from kinetic GISAXS measurements [55]. *Source:* From Papadakis et al. [55]. © 2008, American Chemical Society.

features along the specular reflectivity pattern compared to the patterns in the last two panels in Figure 6.19 was observed [57].

In toluene, the initial film state comprised parallel lamellae, and swelling of the film was observed in the solvent vapour with development of in-plane roughness and ripples over a three- to five-minute timescale, followed by flattening of the swollen film, as illustrated in Figure 6.20 [55].

These processes were deduced from a combination of analysis of the width and position of the diffuse Bragg sheets, as well as the development of side peaks in scans along $q_y$, which signal the presence of in-plane density correlations [55]. Contrasting behaviour was observed during cyclohexane solvent annealing of a symmetric diblock copolymer film with an initial state of partially parallel lamellae [56]. First, a transient state of swollen lamellae with a greater degree of parallel alignment with respect to the substrate was observed, followed (after about 14 minutes) by the evolution of a final disordered state (no lamellar ordering) [56]. Solvent selectivity and volatility will influence the solvent vapour annealing behaviour of block copolymer films. GISAXS has been used by other groups to investigate solvent annealing effects in block copolymer films, see for example Ref. [59–62].

An example of GISANS on a system subjected to dynamic alignment is represented by the data shown in Figure 6.21 for a nanotube-forming peptide under shear in a modified rheometer (Figure 6.21 shows the geometry) [63]. The GISANS measurements themselves are not time-resolved due to the much lower flux of the neutron beam compared to that at a synchrotron GISAXS beamline (in fact the acquisition time per pattern was five hours). However, clear increases in anisotropy reflecting the nematic alignment of the nanotubes under flow is observed at the higher shear rates.

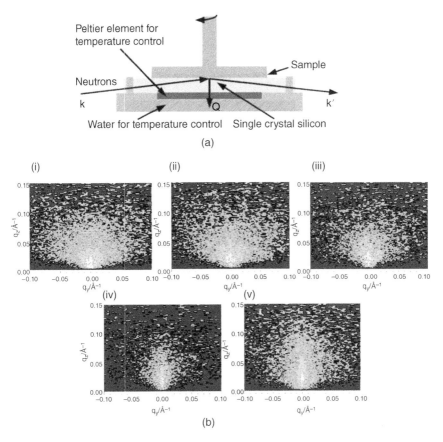

**Figure 6.21** (a) Rheo-GISANS cone-and-plate geometry showing the neutron transmission through a Si plate. (b) GISANS data from a 1 wt% solution of peptide $RFL_4FR$ (letters indicate amino acids) at rest or under steady shear (i) at rest (following shear at a steady shear rate $\dot{\gamma} = 1000\,s^{-1}$), (ii) $\dot{\gamma} = 0.01\,s^{-1}$, (iii) $1\,s^{-1}$, (iv) $100\,s^{-1}$, (v) $1000\,s^{-1}$. The $q_z$ axis is along the shear gradient direction and the $q_y$ direction is mainly oriented along the shear flow direction. The intensity is plotted on a logarithmic colour scale. *Source:* From Hamley et al. [63]. © 2017, American Chemical Society.

Other examples of GISAXS and GISANS measurements on polymers have been reviewed [4, 5].

## 6.7 EXPERIMENTAL EXAMPLES OF GIWAXS/GIXD DATA

Like transmission wide-angle x-ray scattering (WAXS), GIWAXS probes structures on the typical length scale 1–20 Å and has been used to investigate the structure at the surface/in films of nanomaterials, especially nanoparticles, crystalline and ordered polymers, and other small molecules such as those used in optoelectronic materials, e.g. organic solar cells. GIWAXS is also termed grazing-incidence diffraction (GID) and for x-rays, GIXD or grazing-incidence x-ray scattering (GIXS), which are widely used in surface science studies of metal (oxide) surfaces for catalysis. Grazing incidence x-ray diffraction in a surface science context is an important topic in its own right and is the subject of several reviews [9, 64, 65]. This work is outside the scope of this chapter. Here, we give a few examples of GIWAXS/GIXD studies on nanocrystals, lipids, and polymers.

Figure 6.22 shows an example of GISAXS and GIWAXS data for cubic PbSe nanocrystals, which form a rhombohedral superlattice (Figure 6.22b) when deposited on silicon, as determined by GISAXS [66]. The GIWAXS

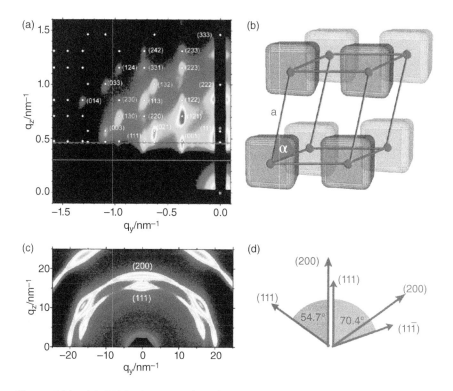

**Figure 6.22** (a) GISAXS pattern for PbSe nanocrystal lattice along with indexed peaks, (b) rhombohedral lattice structure used to index the peaks in (a), (c) GIWAXS image from same sample, (d) showing alignment directions of oriented nanocubes. Red: 'face up', blue: 'corner up'. *Source:* From Choi et al. [66]. © 2012, American Chemical Society.

data (Figure 6.22c) obtained on the same samples (after GISAXS measurements, Figure 6.22b) shows orientation of the cubic PbSe nanocrystals with two coexisting orientations, as shown in Figure 6.22d, one with (100) planes parallel to the substrate and the other with (111) planes in this alignment. The former corresponds to nanocrystals lying 'face up' since the (100) planes form the facets, the latter to their 'corner up' orientation [66].

Grazing-incidence x-ray diffraction has been widely used to investigate the structure of lipid monolayers. In one example, the effect of Hofmeister series anions on the packing of the lipid DPPC [1,2-dipalmitoyl phosphatidylcholine] was investigated by GIXD, among other techniques [67]. Figure 6.23 shows GIXD patterns obtained from monolayers of the lipid at the air-aqueous solution interface obtained by Langmuir-Blodgett

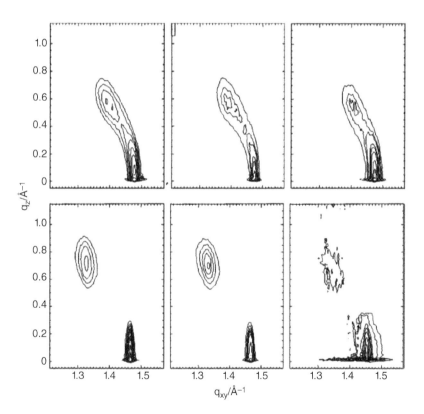

**Figure 6.23** GIXD patterns measured for DPPC monolayers with and without Br⁻ and I⁻ ions. Left: DPPC on water, middle DPPC on 0.5 M NaBr, right: DPPC on 0.5 M NaI. Top: surface pressure $\pi = 45$ mN m$^{-1}$, Bottom: $\pi = 25$ mN m$^{-1}$. Measurements at $\alpha_i = 0.85\ \alpha_c$, where $\alpha_c = 0.13°$. *Source:* From Aroti et al. [67]. © 2004, American Chemical Society.

deposition in the absence and presence of $Br^-$ and $I^-$ ions at two surface pressures. The position and orientation of the peaks provides information on the lipid chain packing (modelled using a rectangular lattice) and tilt angle. At the lower surface pressure ($\pi = 25$ mN m$^{-1}$) the tilt angle of the chains (orientation of the upper off-axis peak in the GIXD pattern with respect to $q_{xy}$) is higher and the lattice is compressed. The peaks are weakened in the presence of NaI.

Examples of kinetic GIWAXS studies include studies on the development of thin film structure during spin casting of bulk heterojunction (BHJ) organic photovoltaic films on solid substrates, with poly-(3-hexylthiophene) (P3HT) as the donor polymer, and [6,6]-phenyl $C_{61}$-butyric acid methyl ester (PCBM) as the small-molecule acceptor [68]. Figure 6.24 shows an example of time-resolved GISAXS and GIWAXS patterns obtained during in situ measurements in the process of spin coating, with fast (0.09 second frames) data acquisition, which is required because the film forms within 8 seconds. GIWAXS probes the formation of P3HT lamellae and GISAXS probes the formation of PCBM aggregates via phase separation. After 8 seconds, the film thickness is stable, and the extents of crystallization and phase separation within the BHJ mixture probed by GIWAXS and GISAXS signals, respectively, become saturated [68]. The development of peaks in the GISAXS and GIWAXS patterns and crystallization and phase separation

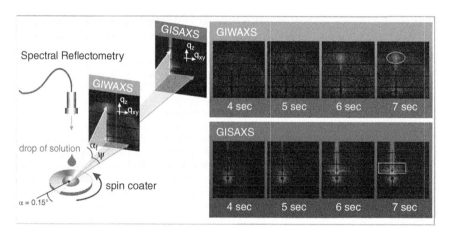

**Figure 6.24** In situ GIWAXS and GISAXS performed during spin coating of an organic photovoltaic polymer film from a chlorobenzene solution. Left: schematic of experiment, right: selected time frames of GIWAXS and GISAXS data. A peak due to lamellar formation is highlighted in an ellipse for the GIWAXS data in the 7 s image, and a box shows diffuse scattering due to PCBM phase separation in the GISAXS data 7 s image. *Source:* From Chou et al. [68].

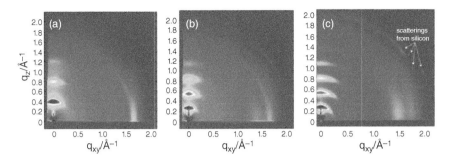

**Figure 6.25** GIWAXS patterns from annealed films of PTzQT semiconducting polymers (a) PTzQT-6, (b) PTzQT-12, (c) PTzQT-14 where the numbers indicate the length of the alkyl chains. *Source:* Figure 3 from Osaka et al. [70].

occur simultaneously. The effect of PCBM content on the phase separation and morphology was also examined in this study, indicating that increased PCBM leads to the loss of lamellar ordering [68].

A similar study also used GIWAXS to probe the in situ structure formation of a different BHJ system during spin coating from chlorobenzene onto a solid substrate, along with the effect of a structuring/plasticizing solvent, di-iodooctane [69]. This system also showed fast film formation kinetics (within 5 seconds in the absence of the di-iodooctane), resolved by rapid GIWAXS measurements.

In another example, GIWAXS (GIXD) provided important information on the extent of lamellar ordering in three thiophene-containing semiconducting polymers bearing thiazolothiazole (PTzQT) units with alkyl chain side groups $C_6H_{13}$, $C_{12}H_{25}$, or $C_{14}H_{29}$ [70]. GIWAXS patterns are shown in Figure 6.25 for films deposited on treated silicon surfaces. The degree of lamellar order (of layers perpendicular to the substrate, cf. schematic in Figure 6.13) increases with alkyl chain side group length due to enhanced layer formation of the polymer (in the centre of the lamellae), driven by the packing of the side chains.

# REFERENCES

1. Renaud, G., Lazzari, R., and Leroy, F. (2009). Probing surface and interface morphology with grazing incidence small angle x-ray scattering. *Surface Science Reports* 64: 255–380.
2. Tolan, M. (1999). *X-ray Scattering from Soft Matter Thin Films*. Berlin: Springer-Verlag.
3. Müller-Buschbaum, P. (2003). Grazing incidence small-angle x-ray scattering: an advanced scattering technique for the investigation of nanostructured polymer films. *Analytical and Bioanalytical Chemistry* 376: 3–10.

4. Müller-Buschbaum, P. (2013). Grazing incidence small-angle neutron scattering: challenges and possibilities. *Polymer Journal* 45: 34–42.

5. Hexemer, A. and Müller-Buschbaum, P. (2015). Advanced grazing-incidence techniques for modern soft-matter materials analysis. *IUCRJ* 2: 106–125.

6. Sinha, S.K. (1991). Reflectivity using neutrons or x-rays - a critical comparison. *Physica B: Condensed Matter* 173: 25–34.

7. Pynn, R. (1992). Neutron-scattering by rough surfaces at grazing-incidence. *Physical Review B* 45: 602–612.

8. Hamley, I.W. and Pedersen, J.S. (1994). Analysis of neutron and x-ray reflectivity. I. Theory. *Journal of Applied Crystallography* 27: 29–35.

9. Als-Nielsen, J., Jacquemain, D., Kjaer, K. et al. (1994). Principles and applications of grazing-incidence x-ray and neutron-scattering from ordered molecular monolayers at the air-water-interface. *Physics Reports-Review Section of Physics Letters* 246: 252–313.

10. Zhou, X.L. and Chen, S.H. (1995). Theoretical foundation of x-ray and neutron reflectometry. *Physics Reports-Review Section of Physics Letters* 257: 223–348.

11. Gibaud, A. and Hazra, S. (2000). X-ray reflectivity and diffuse scattering. *Current Science* 78: 1467–1477.

12. Dosch, H. (1992). *Critical Phenomena at Surfaces and Interfaces. Evanescent X-ray and Neutron Scattering*. Berlin: Springer-Verlag.

13. Higgins, J.S. and Benoît, H.C. (1994). *Polymers and Neutron Scattering*. Oxford: Oxford University Press.

14. Als-Nielsen, J. and McMorrow, D. (2001). *Elements of modern x-ray physics*. Chichester: Wiley.

15. de Jeu, W.H. (2016). *Basic X-ray Scattering for Soft Matter*. Oxford: Oxford University Press.

16. Roe, R.-J. (2000). *Methods of X-ray and Neutron Scattering in Polymer Science*. New York: Oxford University Press.

17. Birkholz, M. (2019). Thin films and multilayers. In: *International Tables for Crystallography, Volume H Powder Diffraction* (eds. C.J. Gilmore, J.A. Kaduk and H. Schenk), 581–600. Wiley.

18. Neutron activation and scattering calculator. (2020). NIST Center for Neutron Research. https://www.ncnr.nist.gov/resources/activation

19. https://sld-calculator.appspot.com/.

20. Russell, T.P. (1990). X-ray and neutron reflectivity for the investigation of polymers. *Materials Science Reports* 5: 171–271.

21. Melnichenko, Y.B. (2016). *Small-Angle Scattering from Confined and Interfacial Fluids*. Berlin: Springer-Verlag.

22. Born, M. and Wolf, E. (1980). *Principles of Optics*, 6th edition. New York: Pergamon Press.

23. Hecht, E. (1987). *Optics*. Reading, MA, USA: Addison-Wesley.

24. Penfold, J. and Thomas, R.K. (1990). The application of the specular reflection of neutrons to the study of surfaces and interfaces. *Journal of Physics, Condensed Matter* 2: 1369–1412.

25. Daillant, J. and Gibaud, A. (eds.) (2009). *X-ray and Neutron Reflectivity Principles and Applications*. Berlin: Springer-Verlag.

26. Pershan, P.S. and Schlossman, M.L. (2012). *Liquid Surfaces and Interfaces. Synchrotron X-ray Methods*. Cambridge, UK: Cambridge University Press.

27. Nevot, L. and Croce, P. (1980). Characterization of surfaces by grazing X-ray reflection - application to study of polishing of some silicate-glasses. *Revue de Physique Appliquee* 15: 761–779.

28. Sinha, S.K., Sirota, E.B., Garoff, S., and Stanley, H.B. (1988). X-ray and neutron-scattering from rough surfaces. *Physical Review B* 38: 2297–2311.

29. Anastasiadis, S.H., Russell, T.P., Satija, S.K., and Majkrzak, C.F. (1989). Neutron reflectivity studies of the surface-induced ordering of diblock copolymer films. *Physical Review Letters* 62: 1852–1855.

30. Penfold, J., Richardson, R.M., Zarbakhsh, A. et al. (1997). Recent advances in the study of chemical surfaces and interfaces by specular neutron reflection. *Journal of the Chemical Society, Faraday Transactions* 93: 3899–3917.
31. Lu, J.R. and Thomas, R.K. (1998). Neutron reflection from wet interfaces. *Journal of the Chemical Society, Faraday Transactions* 94: 995–1018.
32. Bu, W. and Schlossman, M.L. (2016). Synchrotron X-ray scattering from liquid surfaces and interfaces. In: *Synchrotron Light Sources and Free-Electron Lasers* (eds. E. Jaeschke, S. Khan, J. Schneider and J. Hastings), 1579–1616. Cham, Switzerland: Springer.
33. Parratt, L.G. (1954). Surface studies of solids by total Reflection of X-rays. *Physical Review* 95: 359–369.
34. Heavens, O.S. (1955). *Optical Properties of Thin Solid Films*. London: Butterworths Scientific.
35. Azzam, R.M.A. and Bashara, N.M. (1989). *Ellipsometry and Polarized Light*. Amsterdam: North-Holland.
36. Als-Nielsen, J. and Kjaer, K. (1989). X-ray reflectivity and diffraction studies of liquid surfaces and surfactant monolayers. In: *Phase Transitions in Soft Condensed Matter* (eds. T. Riste and D Sherrington), 113–138. New York: Plenum.
37. de Jeu, W.H., Lambooy, P., Hamley, I.W. et al. (1993). On the morphology of a lamellar triblock copolymer film. *Journal de Physique (France) II* 3: 139–146.
38. Vineyard, G.H. (1982). Grazing-incidence diffraction and the distorted-wave approximation for the study of surfaces. *Physical Review B* 26: 4146–4159.
39. Rauscher, M., Salditt, T., and Spohn, H. (1995). Small-angle x-ray scattering under grazing incidence: the cross section in the distorted-wave Born approximation. *Physical Review B* 52: 16855–16863.
40. Salditt, T., Metzger, T.H., Peisl, J. et al. (1995). Determination of the height-height correlation function of rough surfaces from diffuse x-ray scattering. *Europhysics Letters* 32: 331–336.
41. Lazzari, R. (2002). IsGISAXS: a program for grazing-incidence small-angle x-ray scattering analysis of supported islands. *Journal of Applied Crystallography* 35: 406–421.
42. Babonneau, D. (2010). FitGISAXS: software package for modelling and analysis of GISAXS data using IGOR pro. *Journal of Applied Crystallography* 43: 929–936.
43. Senesi, A.J., Eichelsdoerfer, D.J., Macfarlane, R.J. et al. (2013). Stepwise evolution of DNA-programmable nanoparticle superlattices. *Angewandte Chemie, International Edition* 52: 6624–6628.
44. Tate, M.P., Urade, V.N., Kowalski, J.D. et al. (2006). Simulation and interpretation of 2D diffraction patterns from self-assembled nanostructured films at arbitrary angles of incidence: from grazing incidence (above the critical angle) to transmission perpendicular to the substrate. *Journal of Physical Chemistry. B* 110: 9882–9892.
45. Chourou, S.T., Sarje, A., Li, X.Y.S. et al. (2013). HipGISAXS: a high-performance computing code for simulating grazing-incidence x-ray scattering data. *Journal of Applied Crystallography* 46: 1781–1795.
46. Pospelov, G., Van Herck, W., Burle, J. et al. (2020). BornAgain: software for simulating and fitting grazing-incidence small-angle scattering. *Journal of Applied Crystallography* 53: 262–276.
47. Hamley, I.W. (2003). Nanostructure fabrication using block copolymers. *Nanotechnology* 14: R39–R54.
48. Darling, S.B. (2007). Directing the self-assembly of block copolymers. *Progress in Polymer Science* 32: 1152–1204.
49. Hamley, I.W. (2009). Ordering in thin films of block copolymers: fundamentals to potential applications. *Progress in Polymer Science* 34: 1161–1210.
50. Khanna, V., Cochran, E.W., Hexemer, A. et al. (2006). Effect of chain architecture and surface energies on the ordering behavior of lamellar and cylinder forming block copolymers. *Macromolecules* 39: 9346–9356.

51. Disch, S., Wetterskog, E., Hermann, R.P. et al. (2011). Shape induced symmetry in self-assembled mesocrystals of iron oxide nanocubes. *Nano Letters* 11: 1651–1656.

52. Li, T., Senesi, A.J., and Lee, B. (2016). Small angle x-ray scattering for nanoparticle research. *Chemical Reviews* 116: 11128–11180.

53. Rittman, M., Amenitsch, H., Rappolt, M. et al. (2013). Control and analysis of oriented thin films of lipid inverse bicontinuous cubic phases using grazing incidence small-angle x-ray scattering. *Langmuir* 29: 9874–9880.

54. Busch, P., Posselt, D., Smilgies, D.M. et al. (2007). Inner structure of thin films of lamellar poly(styrene-b-butadiene) diblock copolymers as revealed by grazing-incidence small-angle scattering. *Macromolecules* 40: 630–640.

55. Papadakis, C.M., Di, Z.Y., Posselt, D., and Smilgies, D.M. (2008). Structural instabilities in lamellar diblock copolymer thin films during solvent vapor uptake. *Langmuir* 24: 13815–13818.

56. Di, Z.Y., Posselt, D., Smilgies, D.M., and Papadakis, C.M. (2010). Structural rearrangements in a lamellar Diblock copolymer thin film during treatment with saturated solvent vapor. *Macromolecules* 43: 418–427.

57. Zhang, J.Q., Posselt, D., Smilgies, D.M. et al. (2014). Lamellar diblock copolymer thin films during solvent vapor annealing studied by GISAXS: different behavior of parallel and perpendicular lamellae. *Macromolecules* 47: 5711–5718.

58. Busch, P., Rauscher, M., Smilgies, D.M. et al. (2006). Grazing-incidence small-angle x-ray scattering from thin polymer films with lamellar structures-the scattering cross section in the distorted-wave Born approximation. *Journal of Applied Crystallography* 39: 433–442.

59. Cavicchi, K.A., Berthiaume, K.J., and Russell, T.P. (2005). Solvent annealing thin films of poly(isoprene-b-lactide). *Polymer* 46: 11635–11639.

60. Bang, J., Kim, B.J., Stein, G.E. et al. (2007). Effect of humidity on the ordering of PEO-based copolymer thin films. *Macromolecules* 40: 7019–7025.

61. Park, S., Kim, B., Xu, J. et al. (2009). Lateral ordering of cylindrical microdomains under solvent vapor. *Macromolecules* 42: 1278–1284.

62. Paik, M.Y., Bosworth, J.K., Smilges, D.M. et al. (2010). Reversible morphology control in block copolymer films via solvent vapor processing: an in situ GISAXS study. *Macromolecules* 43: 4253–4260.

63. Hamley, I.W., Burholt, S., Hutchinson, J. et al. (2017). Shear alignment of bola-Amphiphilic arginine-coated peptide nanotubes. *Biomacromolecules* 18: 141–149.

64. Robinson, I.K. and Tweet, D.J. (1992). Surface x-ray-diffraction. *Reports on Progress in Physics* 55: 599–651.

65. Dietrich, S. and Haase, A. (1995). Scattering of x-rays and neutrons at interfaces. *Physics Reports-Review Section of Physics Letters* 260: 1–138.

66. Choi, J.J., Bian, K.F., Baumgardner, W.J. et al. (2012). Interface-induced nucleation, orientational alignment and symmetry transformations in nanocube superlattices. *Nano Letters* 12: 4791–4798.

67. Aroti, A., Leontidis, E., Maltseva, E., and Brezesinski, G. (2004). Effects of Hofmeister anions on DPPC Langmuir monolayers at the air-water interface. *Journal of Physical Chemistry. B* 108: 15238–15245.

68. Chou, K.W., Yan, B.Y., Li, R.P. et al. (2013). Spin-cast bulk heterojunction solar cells: a dynamical investigation. *Advanced Materials* 25: 1923–1929.

69. Perez, L.A., Chou, K.W., Love, J.A. et al. (2013). Solvent additive effects on small molecule crystallization in bulk heterojunction solar cells probed during spin casting. *Advanced Materials* 25: 6380–6384.

70. Osaka, I., Zhang, R., Sauve, G. et al. (2009). High-lamellar ordering and amorphous-like $\pi$-network in short-chain thiazolothiazole-thiophene copolymers Lead to high mobilities. *Journal of the American Chemical Society* 131: 2521–2529.

# Index

*Small-Angle Scattering: Theory, Instrumentation, Data and Applications,*
First Edition. Ian W. Hamley.
© 2021 John Wiley & Sons Ltd. Published 2021 by John Wiley & Sons Ltd.